High-Yield Gross Anatomy

High-Yield Gross Anatomy

Ronald W. Dudek, Ph.D

Professor

Department of Anatomy and Cell Biology

East Carolina School of Medicine

Greenville, North Carolina

Williams & Wilkins

A WAVERLY COMPANY

BALTIMORE • PHILADELPHIA • LONDON • PARIS • BANGKOK
BUENOS AIRES • HONG KONG • MUNICH • SYDNEY • TOKYO • WROCLAW

Editor: Elizabeth Nieginski
Manager, Development Editing: Julie A. Scardiglia
Managing Editor: Darrin Kiessling
Marketing Manager: Rebecca Himmelheber
Development Editor: Carol Loyd
Production Coordinator: Felecia R. Weber
Illustration Planner: Felecia R. Weber
Cover Designer: Karen Klinedinst
Typesetter: Maryland Composition
Printer: Port City Press
Digitized Illustrations: Maryland Composition
Binder: Port City Press

Copyright © 1997 Williams & Wilkins

351 West Camden Street
Baltimore, Maryland 21201-2436 USA

Rose Tree Corporate Center
1400 North Providence Road
Building II, Suite 5025
Media, Pennsylvania 19063-2043 USA

Accurate indications, adverse reactions and dosage schedules for drugs are provided in this book, but it is possible that they may change. The reader is urged to review the package information data of the manufacturers of the medications mentioned.

Printed in the United States of America

First Edition

Library of Congress Cataloging-in-Publication Data

Dudek, Ronald W., 1950–
    High-yield gross anatomy / Ronald W. Dudek. — 1st ed.
      p. cm.
    ISBN 0–683–18215–3
    1. Human anatomy—Outlines, syllabi, etc.   I. Title.
    [DNLM: 1. Anatomy—Outlines. 2. Human anatomy—Outlines. syllabi,
etc. QS 18.2 D845h 1997]
QM31.D83   1997
611—dc21
DNLM/DLC   97–6715
for Library of Congress   CIP

The publishers have made every effort to trace the copyright holders for borrowed material. If they have inadvertently overlooked any, they will be pleased to make the necessary arrangements at the first opportunity.

To purchase additional copies of this book, call our customer service department at **(800) 638- 0672** or fax orders to **(800) 447-8438.** For other book services, including chapter reprints and large quantity sales, ask for the Special Sales Department.

Canadian customers should call **(800) 665-1148,** or fax **(800) 665-0103.** For all other calls originating outside of the United States, please call **(410) 528-4223** or fax us at **(410) 528-8550.**

Visit *Williams & Wilkins* on the Internet: http://www.wwilkins.com or contact our customer service department at **custserv@wwilkins.com.** Williams & Wilkins customer service representatives are available from 8:30 am to 6:00 pm, EST, Monday through Friday, for telephone access.

98 99 00

2 3 4 5 6 7 8 9 10

Contents

Preface

High-Yield Gross Anatomy is gross anatomy at its irreducible minimum, and contains many of the recurring themes of the USMLE Step 1. The information presented in *High-Yield Gross Anatomy* prepares you to handle not only the clinical vignettes found on the USMLE, but also the questions that test basic gross anatomy concepts. *High-Yield Gross Anatomy* touches on all the major themes and concepts of gross anatomy, which are presented in a regional, rather than systemic, organization.

Like the USMLE Step 1, the discussions are comprehensively illustrated using a combination of artwork, MRIs and x-rays, including cross-sectional anatomy diagrams. In addition, *High-Yield Gross Anatomy* directly addresses the clinical vignettes of the USMLE Step 1 by incorporating relevant clinical issues that require basic gross anatomy to deduce the correct answer. A number of very common clinical techniques (such as liver biopsy, tracheostomy, and lumbar puncture), which all require a knowledge of the accompanying gross anatomy relationships, are also included.

High-Yield Gross Anatomy, along with *High-Yield Embryology* and *High-Yield Histology*, completes my contribution to the High-Yield series, which is dedicated to improving student performance on the USMLE. In this regard, I would appreciate any comments, suggestions, or additions to any of these review books, especially after you have taken the USMLE. Your input will greatly assist me in future revisions and printings of these titles. You may contact me at dudek@brody.med.ecu.edu.

1

Vertebral Column and Spinal Cord

I. THE VERTEBRAL COLUMN (Figure 1-1). The vertebral column consists of 33 vertebrae [C1–7, T1–12, L1–5, S1–5 (sacrum), and Co1–4 (coccyx)]. The **vertebral canal** contains the spinal cord, dorsal nerve root, ventral nerve root, and meninges. The spinal nerve is located outside the vertebral canal by exiting through the **intervertebral foramen.**

A. Curves

   1. Primary curves are thoracic and sacral curvatures that form during the fetal period.

   2. Secondary curves are cervical and lumbar curvatures that form after birth as a result of lifting the head and walking, respectively.

   3. Kyphosis is an exaggeration of the thoracic curvature, which can occur in the aged due to osteoporosis or disc degeneration.

   4. Lordosis is an exaggeration of the lumbar curvature, which can occur as a result of pregnancy, spondylolisthesis, or a potbelly.

   5. Scoliosis is a complex lateral deviation/torsion that can be caused by poliomyelitis, a short leg, or hip disease.

B. Joints

   1. Atlanto-occipital joints. Nodding the head (as in indicating "yes") occurs at the **atlanto-occipital joints** between C1 (atlas) and the occipital condyles. These joints are synovial and have **no** intervertebral disc. The **anterior** and **posterior atlanto-occipital membranes** limit excessive movement at this joint.

   2. Atlantoaxial joints. Turning the head side to side (as in indicating "no") occurs at the **atlantoaxial joints** between C1 (atlas) and C2 (axis). These are synovial joints and have **no** intervertebral disc. The **alar ligaments** limit excessive movement at this joint.

C. Disorders

   1. Atlantoaxial dislocation

   a. The **rupture of the cruciform (transverse) ligament** due to trauma or rheumatoid arthritis allows mobility of the **dens** (part of the atlas) within the vertebral canal (Figure 1-2). This mobility is called an **atlantoaxial dislocation** and places the cervical spinal cord and medulla at risk.

   b. The **dens** is secured in its position by the cruciform, alar, and apical ligaments and by the tectorial membrane, which is a continuation of the posterior longitudinal ligament.

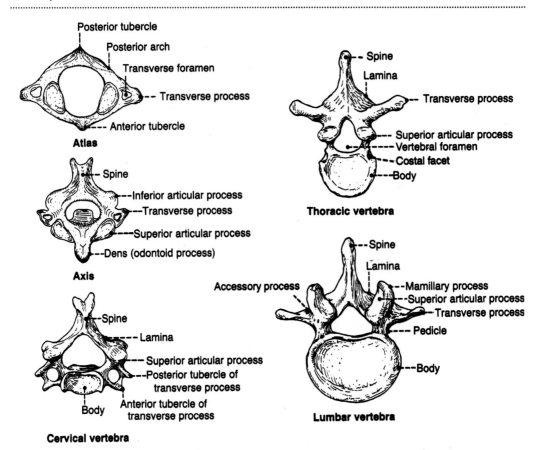

Figure 1-1. Schematic diagram of typical cervical, thoracic, and lumbar vertebrae. Reproduced with permission from Chung, KW: *BRS Gross Anatomy*, 2nd edition, Baltimore, Williams & Wilkins, 1991.

2. Denervation of facet joints

    a. **Facet joints** are synovial joints between **inferior** and **superior articular facets.** These joints are **located** near the intervertebral foramen.

    b. If these joints are **traumatized** or **diseased** (e.g., rheumatoid arthritis), a spinal nerve may be impinged and cause severe pain. To relieve the pain, medial branches of the dorsal primary ramus are severed.

3. Protrusion of the nucleus pulposus

    a. An intervertebral disc consists of the **annulus fibrosus** (fibrocartilage) and **nucleus pulposus** (remnant of the embryonic notochord). The nucleus pulposus generally herniates in a posterior-lateral direction and compresses a nerve root.

    b. **Important features** of a herniated disc at various vertebral levels are shown in Table 1-1.

4. **Dislocations without fracture** occur only in the cervical region because the articular surfaces are inclined horizontally. Cervical dislocations stretch the posterior longitudinal ligament.

5. **Dislocations with fracture** occur in the thoracic and lumbar region because the articular surfaces are inclined vertically.

A

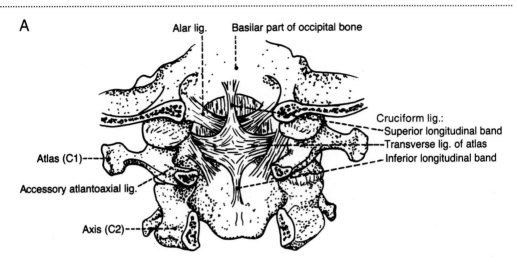

Alar lig.   Basilar part of occipital bone

Cruciform lig.:
- Superior longitudinal band
- Transverse lig. of atlas
- Inferior longitudinal band

Atlas (C1)

Accessory atlantoaxial lig.

Axis (C2)

B

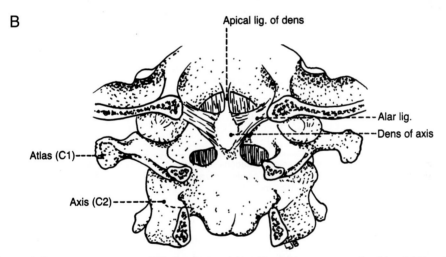

Apical lig. of dens

Alar lig.

Dens of axis

Atlas (C1)

Axis (C2)

Figure 1-2.   A posterior view of the ligaments of the atlas and axis at a superficial level (A) and deeper level (B). The posterior arch and lamina of the atlas and axis have been removed. Reproduced with permission from Chung, KW: *BRS Gross Anatomy,* 2nd edition, Baltimore, Williams & Wilkins, 1991.

6. **Hyperextension of the neck (whiplash)** stretches the anterior longitudinal ligament.

7. A route of **metastasis** for breast, lung, and prostate cancer to the brain exists because the **internal** and **external vertebral venous plexuses** communicate with the cranial dural sinuses and veins of the thorax, abdomen, and pelvis.

8. **Spina bifida occulta** is a common congenital malformation where the **vertebral arch** is absent. The defect is covered by skin and usually is marked by a tuft of hair. This condition is not associated with any neurological deficit.

9. **Hemivertebrae** occurs when a portion of the **vertebral body** fails to develop and can lead to scoliosis.

10. **Sickle cell anemia** is associated with "H-type vertebra" (as observed radiographically) in which central depressions occur in the **vertebral body.**

Table 1-1

Vertebral Levels of a Herniated Disc*

Herniated Disc Between	Compressed Nerve Root	Dermatome Affected	Muscles Affected	Movement Weakness	Reflex Involved
C4 and C5	C5	C5 Shoulder and lateral aspect of arm	Deltoid Biceps	Abduction of arm Flexion of forearm	Biceps jerk
C5 and C6	C6	C6 Lateral arm forearm, and thumb	Extensor carpi radialis longus	Extension of wrist	Biceps jerk
C6 and C7	C7	C7 Posterior arm, forearm, and middle finger	Flexor carpi radialis Triceps	Flexion of wrist Extension of elbow	Triceps jerk
L4 and L5	L5	L5 Lateral thigh, leg, and dorsum of foot	Tibialis anterior Extensor hallucis longus Extensor digitorum longus	Dorsiflexion of ankle Extension of toes	None
L5 and S1	S1	S1 Posterior thigh, leg and lateral part of foot	Gastrocnemius Soleus	Plantar flexion of ankle	Ankle jerk

* This table is not intended to portray real-life clinical situations. Because of the overlap of nerve root contributions to spinal nerves, real-life clinical situations may not be as clear-cut as this table indicates. This table is intended for USMLE review where you are asked to choose the "most likely" answer.

11. **Spondylolisthesis** occurs when the **pedicles** of a lumbar vertebra fail to develop properly. This malformation allows the body of the lumbar vertebra to move anterior with respect to the vertebrae below it, causing a lordosis.

12. **Spondylolysis** is a fracture of the lamina between the inferior and superior articular processes (**pars interarticularis**) within a lumbar vertebra.

13. **Ankylosing spondylitis** is an inflammatory arthritis generally affecting the lumbar vertebrae and sacroiliac joint. The **annulus fibrosus** of the intervertebral discs may become ossified. The ossification bridges the discs at various levels forming a "bamboo spine." A majority of these patients are positive for histocompatibility antigen HLA-B27.

14. **Osteomyelitis** is a bacterial infection that may occur within vertebrae. Tuberculosis and *Staphylococcus aureus* may be causative agents.

D. Reference points

1. **Sacral promontory** is the projecting anterior edge of the S1 vertebral body. It is an important obstetrical landmark.

2. Vertebral levels are used to reference location of important anatomical structures as shown in Table 1-2. Knowledge of these vertebral levels helps when deciphering clinical vignette questions.

II. SPINAL CORD (Figure 1-3)

A. Denticulate ligaments are lateral extensions of **pia mater,** which attach to the dura mater and thereby suspend the spinal cord within the dural sac.

B. Vascular supply (Figure 1-4)

Table 1-2
Vertebral Levels for Reference

Anatomical Structure	Vertebral Level
Hyoid bone Bifurcation of common carotid artery	C4
Thyroid cartilage	C5
Cricoid cartilage Start of trachea Start of esophagus	C6
Sternal notch	T2
Sternal angle Junction of superior and inferior mediastinum Bifurcation of trachea	T4
Pulmonary hilum	T5-T7
IVC hiatus	T8
Xiphisternal joint	T9
Esophageal hiatus	T10
Aortic hiatus	T12
Duodenum	T12-L1
Kidneys	T12-L3
Celiac artery	T12
Superior mesenteric artery Renal artery	L1
End of spinal cord in adult (conus medullaris), pia	L1 or L2
End of spinal cord in new born Inferior mesenteric artery Umbilicus	L3
Iliac crest Bifurcation of aorta	L4
Sacral promontory	S1
End of dural sac, dura, arachnoid, subarachnoid space, and CSF	S2
End of sigmoid colon	S3

IVC = inferior vena cava

CSF = cerebrospinal fluid

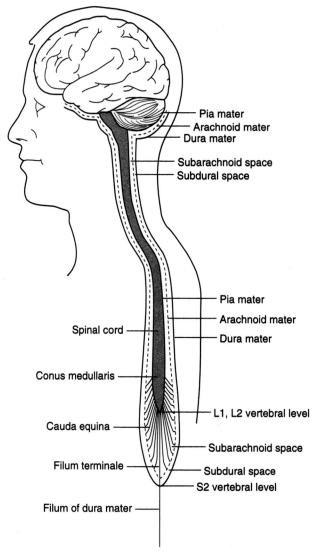

Pia mater
Arachnoid mater
Dura mater

Subarachnoid space
Subdural space

Spinal cord

Pia mater

Arachnoid mater

Dura mater

Conus medullaris

L1, L2 vertebral level

Cauda equina

Subarachnoid space

Filum terminale

Subdural space

S2 vertebral level

Filum of dura mater

Figure 1-3.   A schematic diagram of the spinal cord and meninges. The **subarachnoid space** lies between the pia and arachnoid and contains cerebrospinal fluid (CSF). The **subdural space** is a potential space that lies between the arachnoid and dura. The **epidural space** lies between the bony vertebrae and dura. The epidural space (not shown) contains fat, connective tissue, and the internal vertebral venous plexus. The **cauda equina** is a bundle of nerve roots of lumbar and sacral spinal nerves below the level of termination of the spinal cord. The pia mater as an investing sheet of the spinal cord ends at the conus medullaris. However, the pia mater is reflected onto a fibrous strand called the **filum terminale,** which extends to the end of the dural sac at S2. The filum terminale blends into the **filum of dura mater,** which passes through the sacral canal, exits through the sacral hiatus, and inserts onto the dorsum of the coccyx. Adapted with permission from Chung, KW: *BRS Gross Anatomy,* 2nd edition, Baltimore, Williams & Wilkins, 1991.

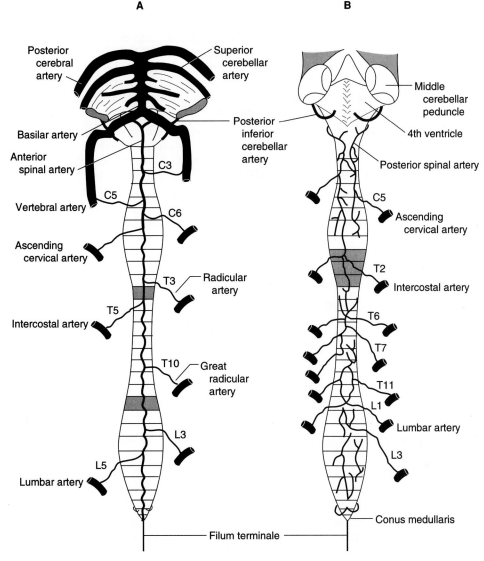

A   B

Posterior cerebral artery

Superior cerebellar artery

Middle cerebellar peduncle

4th ventricle

Posterior inferior cerebellar artery

Basilar artery

Posterior spinal artery

Anterior spinal artery

C3

C5

Vertebral artery

C5

Ascending cervical artery

C6

Ascending cervical artery

T2

Intercostal artery

T3 — Radicular artery

T5

T6

Intercostal artery

T7

T10 — Great radicular artery

T11

L1

L3

Lumbar artery

L5

L3

Lumbar artery

Conus medullaris

Filum terminale

Figure 1-4.   A schematic diagram of the arteries that supply the spinal cord from anterior (A) and posterior (B) views. Radicular arteries are shown at various levels. The great radicular artery is shown branching from a posterior intercostal artery at T10. The shaded areas indicate regions of the spinal cord most vulnerable to arterial blood deprivation. Adapted with permission from Moore, KL: *Clinically Oriented Anatomy*, 3rd edition, Baltimore, Williams & Wilkins, 1992.

1. **Anterior spinal artery** arises from the vertebral arteries and supplies the ventral two thirds of the spinal cord.

2. **Posterior spinal arteries** arise from the vertebral arteries or the posterior inferior cerebellar arteries and supply the dorsal one third of the spinal cord.

3. **Radicular arteries** arise from the vertebral, deep cervical, ascending cervical, posterior intercostal, lumbar, and lateral sacral arteries. The radicular arteries enter the ver-

tebral canal through the intervertebral foramina and branch into the anterior and posterior radicular arteries.

4. Great radicular artery generally arises on the left side from a posterior intercostal artery or a lumbar artery. The great radicular artery is clinically important because it makes a major contribution to the anterior spinal artery and is the main blood supply to the lower part of the spinal cord. If the great radicular artery is ligated during resection of an **aortic aneurysm,** the patient may become paraplegic, impotent, and lose voluntary control of the bladder and bowel.

C. Epidural (caudal) anesthesia is used to relieve pain during childbirth labor. An anesthetic is injected into the **sacral canal** via the **sacral hiatus,** which is marked by the **sacral cornua** (two bony landmarks). The anesthetic diffuses through the dura mater and arachnoid to enter the cerebrospinal fluid (CSF) where it bathes the cauda equina.

D. Lumbar puncture (Figure 1-5) can be done to either withdraw CSF or inject an anesthetic (**spinal block**). A needle is inserted above or below the **spinous process** of the **L4 vertebra.** The needle passes through the following structures:

1. Skin

2. Superficial fascia

3. Supraspinous ligament

4. Interspinous ligament

5. Ligamentum flavum

6. Epidural space containing the internal vertebral venous plexus

7. Dura mater

8. Arachnoid

E. Transection of the spinal cord results in loss of sensation and motor function below the lesion.

1. Paraplegia occurs if the transection occurs anywhere between the cervical and lumbar enlargements of the spinal cord.

2. Quadriplegia occurs if the transection occurs above C3. These individuals may die quickly due to respiratory failure if the phrenic nerve is compromised.

F. Dermatomes (Figure 1-6) are strips of skin extending from the posterior midline to the anterior midline that are supplied by cutaneous branches of dorsal and ventral rami of spinal nerves. A clinical finding of **sensory deficit** in a dermatome is important in order to assess which spinal nerve, nerve root, or spinal cord segment may be damaged. Some dermatomes that are helpful in answering clinical vignette questions are given in Table 1-3.

III. RADIOLOGY

A. **Midsagittal section through the cadaver and MRI image of the cervical region** (Figure 1-7)

B. **Lateral radiograph of the cervical region** (Figure 1-8)

C. **AP radiograph of the lumbar region** (Figure 1-9)

D. **Oblique radiograph of the lumbar region** (Figure 1-10)

E. **Sagittal MRI of the lumbosacral region** (Figure 1-11)

F. **Lateral radiograph of the lumbosacral region** (Figure 1-12)

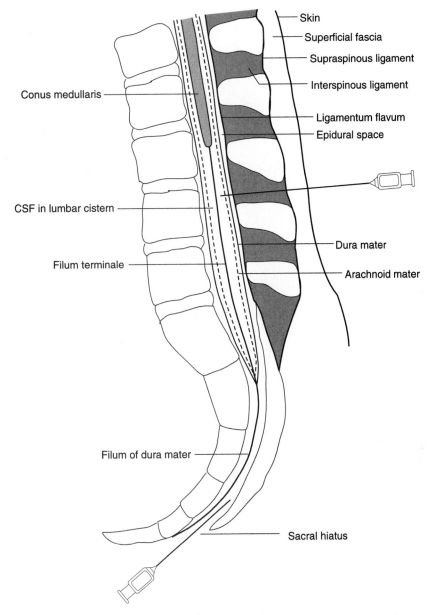

Skin

Superficial fascia

Supraspinous ligament

Interspinous ligament

Conus medullaris

Ligamentum flavum

Epidural space

CSF in lumbar cistern

Dura mater

Filum terminale

Arachnoid mater

Filum of dura mater

Sacral hiatus

Figure 1-5.   A schematic diagram of the lumbar vertebral column and spinal cord. A lumbar puncture needle has been inserted above the spinous process of L4 to withdraw cerebrospinal fluid (CSF). Note the layers the needle must penetrate. Another needle is shown inserted into the sacral canal through the sacral hiatus. This is the site for delivery of an epidural anesthetic. Adapted with permission from Moore, KL: *Clinically Oriented Anatomy,* 3rd edition, Baltimore, Williams & Wilkins, 1992.

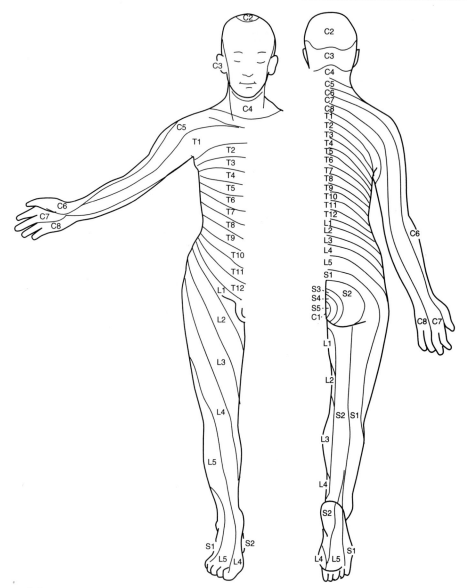

Figure 1-6.   Anterior and posterior views of the dermatomes of the body. Although dermatomes are shown as distinct segments, in reality, there is overlap between any two adjacent dermatomes. Adapted with permission from Chung, KW: *BRS Gross Anatomy*, 2nd edition, Baltimore, Williams & Wilkins, 1991.

Table 1-3

Dermatomes

Structure	Dermatome
Clavicle	C5
Lateral surface of upper limb	C5, C6, C7
Medial surface of upper limb	C8, T1
Thumb	C6
Middle and index fingers	C7
Ring and little fingers	C8
Nipples	T4
Umbilicus	T10
Inguinal region	T12
Anterior and medial surface of lower limb	L1, L2, L3, L4
Foot	L4, L5, S1
Medial side of big toe	L4
Posterior and lateral surface of lower limb	L5, S1, S2
Lateral surface of foot and little toe	S1
Perineum	S2, S3, S4

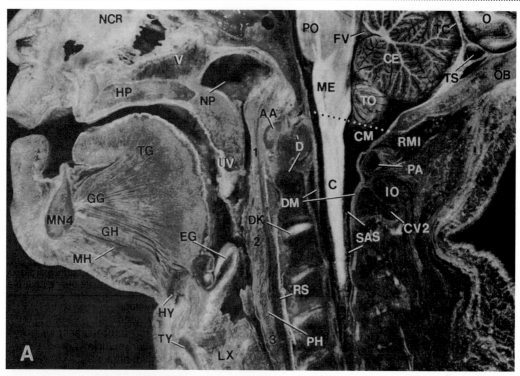

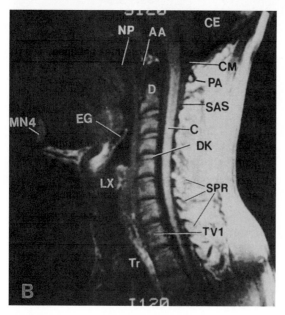

KEY
1 superior pharyngeal con-
   strictor
2 middle pharyngeal constric-
   tor
3 inferior pharyngeal constric-
   tor
AA anterior arch of atlas
C spinal cord
CE cerebellum
CM cisterna magna
CV2 spine, cervical vertebra 2
D dens (odontoid process)
DK intervertebral disk
DM dura mater
EG epiglottis
FV fourth ventricle
GG genioglossus
GH geniohyoid
HP hard palate
HY hyoid bone
IO obliquus capitis inferior
LX larynx (wall)
ME medulla oblongata
MH mylohyoid

MN4 mentis of mandible
NCR nasal cartilage
NP nasopharynx
O occipital lobe
OB occipital bone
PA posterior arch of atlas
PH laryngopharynx
PO pons
RMI rectus capitis minor
RS retropharyngeal space
SAS subarachnoid space
SPR vertebral spinous
   processes
TC tentorium cerebelli
TG tongue
TO tonsil of cerebellum
Tr trachea
TS transverse sinus
TV1 body, thoracic vertebra 1
TY thyroid cartilage
UV uvula
V vomer
White dots level of foramen
   magnum

Figure 1-7.   (A) Midsagittal section through a cadaver and (B) corresponding magnetic resonance image (MRI). Be able to identify the various structures shown. Note in particular the location of the anterior and posterior arches of atlas (C1). Note the relationship of the dens of the axis (C2) to the anterior arch of atlas (C1) and the spinal cord posteriorly. Reproduced with permission from Barrett, CP ; Poliakoff, SJ; Holder, LE; et al.: *Primer of Sectional Anatomy with MRI and CT Correlation,* 2nd edition, Baltimore, Williams & Wilkins, 1994.

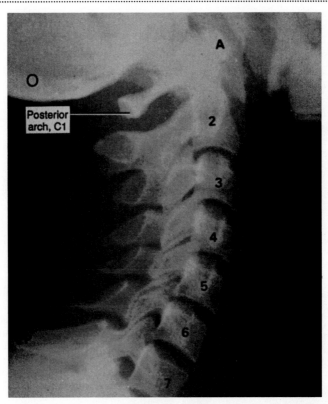

Figure 1-8.   A lateral radiograph of the cervical region showing the cervical vertebrae. The vertebral bodies of C2–C7 are numbered accordingly. Note the location of the anterior (A) and posterior arches of atlas (C1). Observe the vertebra prominens of C7 (VP), occipital bone (O), and angle of the mandible (arrowheads). Reproduced with permission from Moore, KL: *Clinically Oriented Anatomy*, 3rd edition, Baltimore, Williams & Wilkins, 1992.

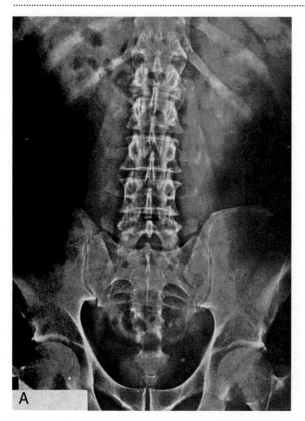

A

Figure 1-9. (A) Anteroposterior (AP) radiograph of the lumbar region. (B) A schematic diagram of the appearance of a lumbar vertebra on an AP radiograph. You should be able to identify the various parts of the lumbar vertebra shown in the schematic on the actual radiograph. Reproduced with permission from Slaby, F and Jacobs, ER: *Radiographic Anatomy,* Baltimore, Williams & Wilkins, 1990.

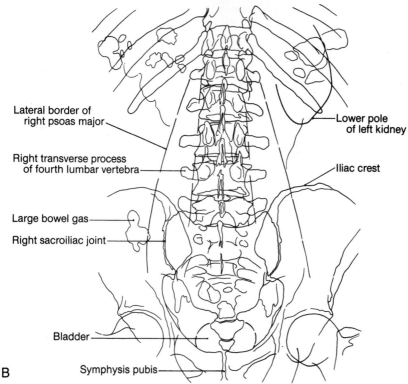

Lateral border of right psoas major

Right transverse process of fourth lumbar vertebra

Large bowel gas

Right sacroiliac joint

Lower pole of left kidney

Iliac crest

Bladder

Symphysis pubis

B

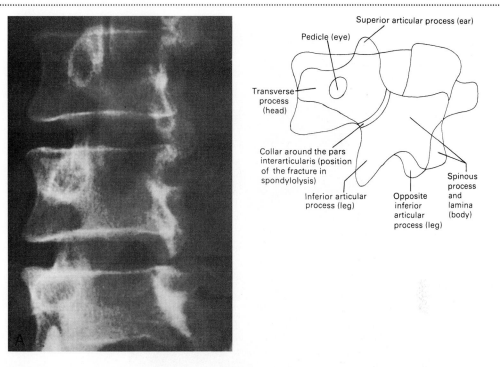

Figure 1-10.   (A) Oblique radiograph of the lumbar region. (B) A schematic diagram of the appearance of a lumbar vertebra on an oblique radiograph showing the characteristic "Scottie dog" appearance. You should be able to identify the various parts of the lumbar vertebra shown in the schematic on the actual radiograph. Reproduced with permission from Pegington, J: *Clinical Anatomy in Action,* Volume 1, Edinburgh, UK, Churchill Livingstone, 1985.

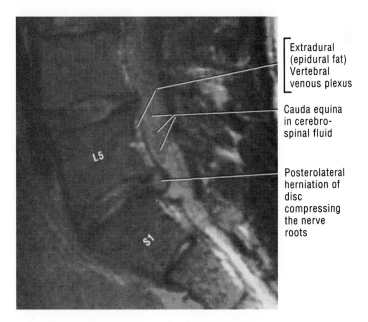

Figure 1-11.   Sagittal magnetic resonance image (MRI) of the lumbosacral region showing the protrusion of the nucleus pulposus through the annulus fibrosus (a herniated disc). Reproduced with permission from Moore, KL: *Clinically Oriented Anatomy,* 3rd edition, Baltimore, Williams & Wilkins, 1992.

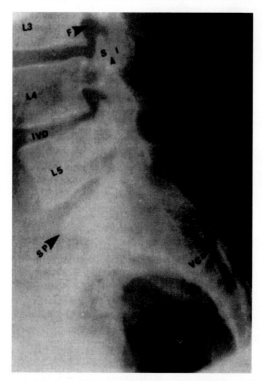

Figure 1-12.   Lateral radiograph of the lumbosacral region. Note the lumbar vertebrae (L3, L4, L5) and the intervertebral discs (*IVD*) between them. The sacral promontory (*SP*) is the anterior edge of the S1 vertebral body. The facet (synovial) joint (small arrowhead) between the inferior articular process of L3 (*I*) and the superior articular process of L4 (*S*) is shown. Note its close proximity to the intervertebral foramen (*F*) through which a spinal nerve exists. Inflammation of the facet joints (arthritis) can involve the spinal nerves. (*VC* = vertebral canal.) Reproduced with permission from Moore, KL: *Clinically Oriented Anatomy*, 3rd edition, Baltimore, Williams & Wilkins, 1992.

2

Thorax

I. BREAST

A. The breast is located in the superficial fascia of the anterior chest wall overlying the **pectoralis major** and **serratus anterior muscles** and extends into the axilla (axillary tail; superior lateral quadrant), where a high percentage of tumors occur.

B. In most women, the breast extends vertically from **rib 2** to **rib 6** and laterally from the **sternum** to the **midaxillary line.**

C. The **retromammary space** lies between the breast and the **pectoral (deep) fascia** and allows free movement of the breast. If **breast carcinoma** invades the retromammary space and pectoral fascia, contraction of the pectoralis major may cause the **whole breast to move superiorly.**

D. **Suspensory (Cooper's) ligaments** extend from the dermis of the skin to the pectoral fascia and provide support for the breast. If **breast carcinoma** invades the suspensory ligaments, these ligaments may shorten and cause **dimpling of the skin** or **inversion of the nipple.**

E. **Adipose tissue** within the breast contributes largely to the contour and size of the breast.

F. **Glandular tissue (mammary gland)** within the breast is a modified sweat gland consisting of acini, which are ultimately drained by 15–20 **lactiferous ducts** that open onto the nipple. Just deep to the surface of the nipple, each lactiferous duct expands into a **lactiferous sinus,** which serves as a reservoir for milk during lactation.

G. **Arterial supply** is from the **internal thoracic, lateral thoracic,** and **intercostal arteries.**

H. The chief **venous drainage** is to the **axillary vein. Internal thoracic, lateral thoracic,** and **intercostal veins** also participate. **Metastasis** of breast carcinoma to the brain may occur by the following route: Cancer cells enter an intercostal vein → vertebral venous plexuses → cranial dural sinuses.

I. The chief **lymphatic drainage** is to the **axillary nodes. Parasternal, clavicular,** and **inguinal nodes** also participate. **Breast carcinoma** may metastasize via the lymphatic vessels or block lymph flow, causing a **thick leathery skin.**

J. Innervation is via **intercostal nerves 2–6 (T2, T3, T4, T5, T6 dermatomes).**

K. During a mastectomy, the **long thoracic nerve** must be preserved. Damage to the long thoracic nerve paralyzes the serratus anterior muscle and causes a "winged scapula."

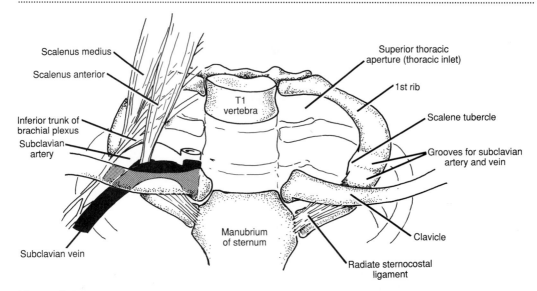

Figure 2-1.   The first pair of ribs and their costal cartilages are shown with their articulation with the T1 vertebra and manubrium of the sternum. On the right, structures crossing rib 1 are shown (subclavian vein, subclavian artery, brachial plexus). Note the relationship of these structures to the clavicle. Reproduced with permission from Moore, KL: *Clinically Oriented Anatomy*, 3rd edition, Baltimore, Williams & Wilkins, 1992.

II.   CHEST WALL (Figure 2-1)

A. Insertion of a right subclavian venous catheter. In this procedure, a needle is introduced at the lower border of the clavicle just lateral to the **midclavicular line.**

   1. The needle is angled toward the **sternoclavicular joint (T2)** and penetrates the following locations:

      a. Skin

      b. Superficial fascia

      c. Subclavius muscle

      d. Clavipectoral fascia

   2. The **anterior scalene muscle** and **right subclavian artery** are located **posterior** to the right subclavian vein in this area. **Improper insertion** of catheter may tear the subclavian vein and/or subclavian artery, resulting in accumulation of blood in the right pleural space (**hemothorax**).

B. Thoracostomy (Figure 2-2) may be necessary to drain fluid out of the pleural cavity by inserting a needle through **intercostal space 4 in the midaxillary line.**

   1. The needle penetrates the following:

      a. Skin

      b. Superficial fascia

      c. Serratus anterior muscle

      d. External intercostal muscle

      e. Internal intercostal muscle

      f. Innermost intercostal muscle

      g. Parietal pleura

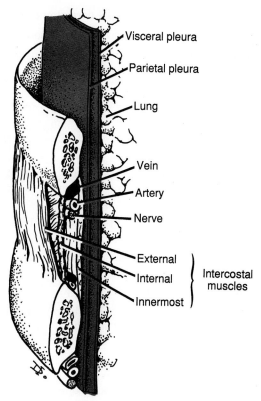

Visceral pleura

Parietal pleura

Lung

Vein

Artery

Nerve

External
Internal } Intercostal muscles
Innermost

Figure 2-2. A schematic diagram of an intercostal space and its relationship to pleura and lung. Reproduced with permission from Moore, KL: *Clinically Oriented Anatomy*, 3rd edition, Baltimore, Williams & Wilkins, 1992.

2. The needle should be inserted close to the **upper** border of the rib to avoid the **intercostal vein, artery,** and **nerve (VAN),** which run in the costal groove between the internal intercostal muscle and innermost intercostal muscle.

C. Intercostal nerve block may be necessary to relieve pain associated with a rib fracture or herpes zoster (shingles). A needle is inserted at the posterior angle of the rib along the **lower** border of the rib in order to bathe the nerve in anesthetic.

    1. Several intercostal nerves must be blocked to achieve pain relief because of the presence of nerve collaterals [i.e., overlapping of contiguous **dermatomes** (see Chapter 1 II F)].

    2. Supraclavicular nerves (C4 and **C5 dermatomes)** supply the chest wall above the sternal angle.

    3. Intercostal nerves 1 and **2 (T1** and **T2 dermatomes)** join the brachial plexus and, therefore, supply the chest wall and upper limb.

    4. Intercostal nerves 3, 4, 5, 6 (T3, T4, T5, T6 dermatomes) supply the chest wall.

    5. Intercostal nerves 7, 8, 9, 10, 11 (T7, T8, T9, T10, T11 dermatomes) supply the chest wall and abdominal wall.

    6. Subcostal nerve (T12 dermatome) follows the lower border of rib 12 and supplies the abdominal wall and upper inguinal region.

7. The level of the **nipple** corresponds to **T4 dermatome.**

8. The level of the **umbilicus** corresponds to **T10 dermatome.**

D. Aneurysm of the aorta may appear as a pulsatile swelling in the sternal notch (T2) because the arch of the aorta lies behind the manubrium.

E. Coarctation (constriction) of the aorta is a congenital malformation associated with increased blood pressure to the upper extremities, lack of femoral artery pulse, and high risk of cerebral hemorrhage and bacterial endocarditis.

  1. Coarctation of the aorta generally is located distal to the **left** subclavian artery and inferior to the ligamentum arteriosum.

  2. The **internal thoracic, intercostal, superior epigastric, inferior epigastric,** and **external iliac arteries** are involved in the collateral circulation to bypass the constriction and become dilated.

  3. Dilation of the intercostal arteries causes erosion of the lower border of the ribs (**rib notching**).

F. Knife wound to chest wall above the clavicle may damage structures at the root of the neck.

  1. The **subclavian artery** may be cut.

  2. The **lower trunk of the brachial plexus** may be cut, causing loss of hand movements (ulnar nerve involvement) and loss of sensation over the medial aspect of the arm, forearm, and last two digits (C8 and T1 dermatomes).

  3. The **cervical pleura** and **apex of the lung** may be cut, causing an open pneumothorax and collapse of the lung. These structures project superiorly into the neck through the thoracic inlet and posterior to the sternocleidomastoid muscle.

G. Fractures of the lower ribs (Figure 2-3) may damage abdominal viscera.

  1. On the right side, the **right kidney** and **liver** may be damaged.

  2. On the left side, the **left kidney** and **spleen** may be damaged.

  3. On both sides, the **pleura** may be damaged because it extends down to rib 12 just lateral to the vertebrae (**costodiaphragmatic recess**).

H. Projections of the diaphragm on the chest wall

  1. The **central tendon** of the diaphragm lies directly posterior to the xiphosternal joint.

  2. The **right dome** of the diaphragm arches superiorly to the **upper** border of rib 5 in the midclavicular line.

  3. The **left dome** of the diaphragm arches superiorly to the **lower** border of rib 5 in the midclavicular line.

III. PLEURA, TRACHEOBRONCHIAL TREE, AND LUNGS

A. Pleura

  1. Visceral pleura adheres to the lung on all its surfaces. The visceral and parietal pleura are continuous at the root of the lung.

  2. Parietal pleura adheres to the chest wall, diaphragm, and pericardial sac.

  a. The parietal pleura is **named according to the anatomical region** with which it is associated.

  (1) Costal pleura is associated with the internal surface of the sternum, costal cartilages, ribs, and sides of the thoracic vertebrae.

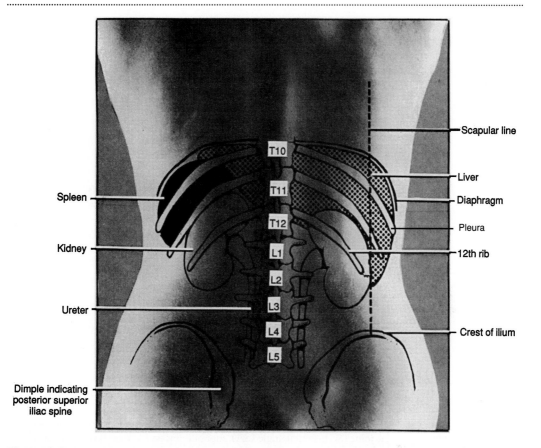

Scapular line

Liver

Diaphragm

Pleura

12th rib

Crest of ilium

Spleen

Kidney

Ureter

Dimple indicating
posterior superior
iliac spine

Figure 2-3.   Surface anatomy and markings on the back of a 21-year-old woman. Note that the kidneys are located from T12 to L3 vertebrae and that the right kidney is lower than the left. Note the pleura extending from T12, across rib 12, to rib 10. Also, review the structures that may be injured by fractures to the lower ribs. Reproduced with permission from Moore, KL: *Clinically Oriented Anatomy*, 3rd edition, Baltimore, Williams & Wilkins, 1992.

(2) **Mediastinal pleura** is associated with the **mediastinum** and forms the **pulmonary ligament** (located inferior to the root of the lung), which serves to support the lung.

(3) **Diaphragmatic pleura** is associated with the diaphragm.

(4) **Cervical pleura** is associated with the root of the neck.

b. The **boundaries** of parietal pleura (pleural reflections) are important clinical landmarks (Figure 2-4).

(1) **Right-side reflection** extends from the:

(a) Sternoclavicular joint to the anterior midline at the sternal angle (T4), where the two pleural sacs come into contact and may overlap

(b) Inferiorly to the xiphoid process, where the pleura extends beyond the rib cage at the infrasternal notch

(c) Laterally across rib 8 at the midclavicular line and rib 10 at the midaxillary line

(d) Posteriorly across rib 12 at its neck

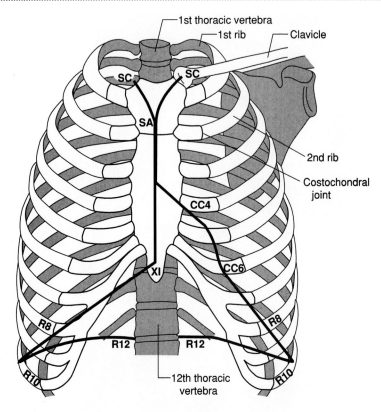

Figure 2-4. A schematic diagram indicating the pleural lines of reflection. On the right side, the pleura extends from the sternoclavicular joint (SC), sternal angle (SA), xiphoid process (XI), rib 8 (R8), rib 10 (R10), and rib 12 (R12). On the left side, the pleura extends from the sternoclavicular joint (SC), sternal angle (SA), costal cartilage 4 and 6, rib 8 (R8), rib 10 (R10), and rib 12 (R12). Adapted with permission from Moore, KL: *Clinically Oriented Anatomy*, 3rd edition, Baltimore, Williams & Wilkins, 1992.

(2) **Left-side reflection** extends from the:

(a) Sternoclavicular joint to the anterior midline at the sternal angle (T4), where the two pleural sacs come into contact and may overlap

(b) Inferiorly to costal cartilage 4 and laterally to costal cartilage 6 (cardiac notch)

(c) Laterally across rib 8 at the midclavicular line and rib 10 at the midaxillary line

(d) Posteriorly across rib 12 at its neck

3. **Pleural recesses**

a. Right and **left costodiaphragmatic recesses** are slit-like spaces between the costal and diaphragmatic parietal pleura.

(1) **During inspiration,** the lungs descend into the right and left costodiaphragmatic recesses, causing the recesses to appear radiolucent (dark) on x-ray film.

(2) **During expiration,** the lungs ascend so that the costal and diaphragmatic parietal pleura come together and the radiolucency disappears on x-ray film.

(3) The **costodiaphragmatic (costophrenic) angle** should appear sharp in a PA x-ray film. If the angle is blunted, suspected pathology of the pleural space may be excess fluid, blood, tumor, or scar tissue.

(4) With a patient in the standing position, **excess fluid** within the pleural cavity will accumulate in the costodiaphragmatic recesses.

b. **Right** and **left costomediastinal recesses** are slit-like spaces between the costal and mediastinal parietal pleura.

(1) **During inspiration,** the anterior borders of both lungs expand and enter the right and left costomediastinal recesses. In addition, the **lingula of the left lung** expands and enters a portion of the **left** costomediastinal recess, causing that portion of the recess to appear radiolucent (dark) on x-ray film.

(2) **During expiration,** the anterior borders of both lungs recede and exit the right and left costomediastinal recesses.

4. **Pleuritis** is inflammation of the pleura.

a. Pleuritis involving only the **visceral pleura** is **not** associated with pain because the visceral pleura receives no nerve fibers of general sensation.

b. Pleuritis involving the **parietal pleura** is associated with sharp local pain and referred pain to the thoracic and abdominal wall because parietal pleura is innervated by **intercostal nerves.** Referred pain also may be felt at the root of the neck and the shoulder because the mediastinal and central part of diaphragmatic parietal pleura are innervated by the **phrenic nerve** (C3, C4, C5).

5. **Inadvertent damage** to the pleura may occur during a:

a. **Surgical posterior approach to the kidney.** If rib 12 is very short, rib 12 may be mistaken for rib 11. An incision prolonged to the level of rib 11 damages the pleura.

b. **Abdominal incision at the right infrasternal angle.** The pleura extends beyond the rib cage in this area.

c. **Stellate ganglion nerve block**

d. **Brachial plexus nerve block**

e. **Knife wounds to the chest wall above the clavicle**

6. **Spontaneous pneumothorax** occurs when air enters the pleural cavity usually due to a ruptured bleb (bullus) of a diseased lung.

7. **Open pneumothorax** occurs when the parietal pleura is pierced (e.g., knife wound), and the pleural cavity is opened to the outside atmosphere. Upon inspiration, air is sucked into the pleural cavity.

8. **Tension pneumothorax** may occur as a sequela to an open pneumothorax if the inspired air cannot leave the pleural cavity through the wound upon expiration. As a result, there is a collapsed lung on the wounded side and a compressed lung on the opposite side due to a deflected mediastinum.

B. **Tracheobronchial tree**

1. The trachea is a tube composed of **16–20 U-shaped hyaline cartilages** and the **trachealis** muscle.

2. The **trachea** begins just inferior to the cricoid cartilage (C6 vertebra) and ends at the sternal angle (T4 vertebra), where it bifurcates into the **right** and **left main bronchi.**

3. At the bifurcation of the trachea, the last tracheal cartilage forms the **carina,** which can be observed by bronchoscopy as a raised ridge of tissue in the **sagittal plane.**

4. The **right main bronchus** is shorter, wider, and turns to the right at a shallower angle than the left main bronchus. The right main bronchus branches into 3 **lobar bronchi** (upper, middle, lower) and finally into **10 segmental bronchi.**

5. The **left main bronchus** branches into **2 lobar bronchi** (upper, lower) and finally into **8–10 segmental bronchi.**

6. Branching of segmental bronchi corresponds to the **bronchopulmonary segments** of the lung and is particularly variable within the **lower lobes** of both lungs.

7. **Distortions in the position of the carina** may indicate **metastasis of bronchogenic** carcinoma into the tracheobronchial lymph nodes that surround the tracheal bifurcation or **enlargement of the left atrium.** The mucous membrane covering the carina is very sensitive in eliciting the **cough reflex.**

8. **Compression of the trachea** may be caused by an **enlargement of the thyroid gland** or an **aortic arch aneurysm.** The aortic arch aneurysm tugs on the trachea with each cardiac systole and can be felt by palpating the trachea at the sternal notch.

9. **Aspiration of foreign objects** generally occurs in the **right main bronchus, right middle lobar bronchus,** or **right lower lobar bronchus.** Asymmetric lung volumes may be found on a PA x-ray film if air can get past the obstruction during inspiration but cannot get out during expiration.

C. Lungs

1. Right lung

a. The right lung consists of **3 lobes** (upper, middle, lower) separated by a **horizontal fissure** and an **oblique fissure.**

(1) The **right upper** lobe lies in a superior/anterior position.

(2) The **right middle** lobe lies in an anterior position between costal cartilages 4 and 6.

(3) The **right lower** lobe lies in an inferior/posterior position.

b. The **horizontal fissure** runs at the level of **costal cartilage 4** and meets the **oblique fissure** at the midaxillary line.

c. The **diaphragmatic surface** consists of the middle and lower lobes.

2. Left lung

a. The left lung consists of **2 lobes** (upper, lower) separated by an **oblique fissure.** The **upper lobe** contains the **cardiac notch** where the left ventricle and pericardial sac abut the lung. Just beneath the cardiac notch lies the **lingula,** which is the embryological counterpart to the right middle lobe.

b. The **left upper** lobe lies in a superior/anterior position. The **left lower** lobe lies in an inferior/posterior position.

c. The **diaphragmatic surface** consists of the lower lobe.

3. A **bronchopulmonary segment** contains:

a. A **segmental bronchus, a branch of the pulmonary artery,** and a **branch of the bronchial artery,** which run together

b. **Tributaries of the pulmonary vein,** which are found at the periphery between two adjacent bronchopulmonary segments. These veins form surgical landmarks during segmental resection of the lung.

4. **Atelectasis** is the collapse of alveoli so that they cannot be inflated during inspiration. Atelectasis may occur as a result of:

    a. Inflammation of alveoli, leading to a reduction of surfactant

    b. Tumor or mucous obstruction of a segmental bronchus, leading to **segmental atelectasis.** If a segmental atelectasis occurs, the heart and mediastinum are shifted towards the collapsed side.

5. **Bronchiectasis** is the abnormal dilation of bronchi resulting from damage to the bronchial wall.

6. **Emphysema** and **pulmonary fibrosis** may cause a destruction of **lung elasticity** so that the lungs are unable to recoil adequately, causing incomplete expiration. Consequently, the rectus abdominis, external oblique, internal oblique, transversus abdominis, internal intercostal, and serratus posterior inferior muscles (expiratory muscles) must assist in expiration. Chronic emphysema may be associated with an increased retrosternal air space, increased anterior–posterior diameter of the chest wall, and flattening of the diaphragm.

7. **Silicosis, cancer,** and **pneumonia** may cause a destruction of **lung distensibility** so that the lungs are unable to expand fully upon inspiration. Consequently, the diaphragm, external intercostal, sternocleidomastoid, levator scapulae, serratus anterior, scalenes, pectoralis major and minor, erector spinae, and serratus posterior superior muscles (inspiratory muscles) must work harder to inflate the lungs.

8. **Pulmonary thromboembolism** involves pulmonary artery obstruction by a displaced thrombus possibly from the lower limb (great saphenous vein). If the right pulmonary artery is obstructed, blood is diverted to the left side and causes distention of the left pulmonary artery.

9. **Bronchogenic carcinoma**

    a. Bronchogenic carcinoma tends to affect the upper lobes (anterior aspect) and may involve:

        (1) The **phrenic nerve,** resulting in paralysis of diaphragm on one side

        (2) The **recurrent laryngeal nerve** located near the apex of the lung, resulting in paralysis of the vocal cord causing hoarseness of the voice

        (3) The **superior thoracic nerve, sympathetic chain,** and **stellate ganglion** located near the apex of the lung, resulting in pain over the shoulder and axilla

    b. **Metastasis**

        (1) **Tracheobronchial (mediastinal), parasternal,** and **supraclavicular** lymph nodes are involved in the **lymphatic metastasis** of cancer cells.

            (a) Enlargement of the **mediastinal nodes** may indent the esophagus, which can be observed radiologically during a barium swallow.

            (b) Enlargement of the **supraclavicular nodes** indicates the possibility of malignant disease in thoracic or abdominal organs.

        (2) Metastasis to **brain via arterial blood** may occur by the following route: Cancer cells enter a lung capillary → pulmonary vein → left atrium and ventricle → aorta → internal carotid and vertebral arteries.

        (3) Metastasis to **brain via venous blood** may occur by the following route: Cancer cells enter a bronchial vein → azygous vein → vertebral venous plexuses → cranial dural sinuses.

10. **Reactivation tuberculosis** tends to affect the apices of the lungs.

11. **Miliary tuberculosis** is scattered uniformly throughout the lungs as multiple fine nodules.

12. **Lung abscess** tends to affect the lower lobes.

IV. PERICARDIUM AND HEART

A. Pericardium

1. The pericardium consists of **three layers:**

a. A **visceral** layer of serous pericardium (also called epicardium histologically)

b. A **parietal** layer of serous pericardium

c. A thick connective tissue layer called the **fibrous pericardium**

2. The **pericardial cavity,** which normally contains a small amount of fluid, lies between the visceral layer and parietal layer of serous pericardium.

3. The **fibrous pericardium fuses superiorly** to the adventitia of the great vessels, **inferiorly** to the central tendon of the diaphragm, and **anteriorly** to the sternum.

4. The **phrenic nerve** and **pericardiophrenic artery** descend through the mediastinum lateral to the fibrous pericardium and may be injured during surgery to the heart.

5. The **transverse sinus** is a recess of the pericardial cavity. After the pericardial sac is opened, a surgeon can pass a finger or ligature (from one side of the heart to the other) through the transverse sinus between the great arteries and pulmonary veins.

6. The **oblique sinus** is a recess of the pericardial cavity that ends in a cul-de- sac surrounded by the pulmonary veins.

7. The thoracic portion of the **inferior vena cava (IVC)** lies within the pericardium so that the pericardium must be opened to expose this portion of the IVC.

8. **Cardiac tamponade** is the accumulation of fluid within the pericardial cavity resulting in compression of the heart because the fibrous pericardium is inelastic. Compression of the **superior vena cava (SVC)** may cause the veins of the face and neck to engorge with blood.

9. **Pericardiocentesis,** the removal of fluid from the pericardial cavity, can be approached in two ways:

a. Sternal approach

(1) **Procedure.** A needle is inserted at intercostal space 5 or 6 on the left side near the sternum. The cardiac notch of the left lung leaves the fibrous pericardium exposed at this site.

(2) The **needle** penetrates the following **layers:**

(a) Skin

(b) Superficial fascia

(c) Pectoralis major muscle

(d) External intercostal membrane

(e) Internal intercostal muscle

(f) Transverse thoracic muscle

(g) Fibrous pericardium

(h) Parietal layer of serous pericardium

(3) **Risk.** The internal thoracic artery, coronary arteries, and pleura are in jeopardy during this approach.

b. Subxiphoid approach

(1) **Procedure.** The needle is inserted at the left infrasternal angle in a superior and posterior position.

    (2) The **needle** penetrates the following **layers:**

        (a) Skin

        (b) Superficial fascia

        (c) Anterior rectus sheath

        (d) Rectus abdominis muscle

        (e) Transverse abdominis muscle

        (f) Fibrous pericardium

        (g) Parietal layer of serous pericardium

    (3) Risk. If the needle is not angled properly, the diaphragm and liver may be penetrated.

B. Heart (Figure 2-5)

    1. Surfaces (Table 2-1)

    2. Borders (Table 2-2)

    3. Blood supply. Coronary artery occlusion occurs **most commonly** in the circumflex artery, followed by the right coronary artery, and then the anterior interventricular artery (Table 2-3)**.**

    4. Venous drainage (Table 2-4)

    5. Enlargement of the heart may be caused by pulmonary edema, where blood can back up into the SVC, right atrium, and right ventricle. Such a patient is placed in the upright position when eating because the esophagus lies posterior to the heart.

    6. Auscultation sites (Figure 2-6)

        a. Pulmonary valve—edge of sternum at left intercostal space 2

        b. Aortic valve—edge of sternum at right intercostal space 2

        c. Mitral valve—near cardiac apex at left intercostal space 5

        d. Tricuspid valve—over the sternum at intercostal space 5

V. CROSS-SECTIONAL ANATOMY. When studying the cross sections, note the anterior–posterior position of the various structures. A clinical vignette question may describe a bullet or knife wound at a specific vertebral level and ask which structures would be penetrated in an anterior–posterior sequence.

    A. At T2-T3 where three branches of aortic arch originate (Figure 2-7)

    B. At T6 where the aorticopulmonary window is located (Figure 2-8)

    C. At T7 where the right pulmonary artery originates (Figure 2-9)

    D. At T8 where the ascending aorta and pulmonary trunk originate (Figure 2- 10)

    E. At T10 where inferior vena cava enters right atrium (Figure 2-11)

VI. RADIOLOGY

    A. Posterior–anterior (PA) chest x-ray (Figure 2-12)

    B. A lateral chest x-ray (Figure 2-13)

    C. An aortic angiogram (Figure 2-14)

    D. Right and left coronary arteriogram (Figure 2-15)

A

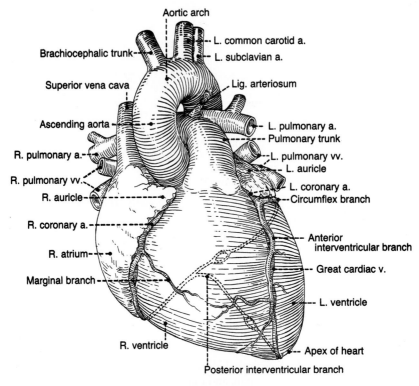

Aortic arch

L. common carotid a.

Brachiocephalic trunk

L. subclavian a.

Superior vena cava

Lig. arteriosum

Ascending aorta

L. pulmonary a.

Pulmonary trunk

R. pulmonary a.

L. pulmonary vv.

R. pulmonary vv.

L. auricle

R. auricle

L. coronary a.

Circumflex branch

R. coronary a.

Anterior interventricular branch

R. atrium

Great cardiac v.

Marginal branch

L. ventricle

R. ventricle

Apex of heart

Posterior interventricular branch

B

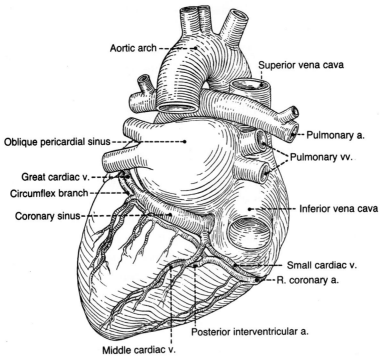

Aortic arch

Superior vena cava

Oblique pericardial sinus

Pulmonary a.

Pulmonary vv.

Great cardiac v.

Circumflex branch

Coronary sinus

Inferior vena cava

Small cardiac v.

R. coronary a.

Posterior interventricular a.

Middle cardiac v.

Figure 2-5. (A) Anterior view of the heart with coronary arteries. (B) Posterior view of the heart. Reproduced with permission from Chung, KW: *BRS Gross Anatomy,* 2nd edition, Baltimore, Williams & Wilkins, 1991.

Table 2-1.
Heart Surfaces

Surface	Structure
Base or posterior surface	Left atrium
Apex	Left ventricle located at intercostal space 5 at the midclavicular line
Sternal surface	Right ventricle
Diaphragmatic surface	Left ventricle

Table 2-2.
Heart Borders

Border	Structure
Right border	Right atrium
Left Border	Left ventricle, left auricle, pulmonary trunk aortic arch
Inferior border	Right ventricle
Superior border	Superior vena cava, aorta, pulmonary trunk

Table 2-3.
Blood Supply to the Heart

	Branches		Structures Supplied
Left coronary artery	Circumflex artery Anterior interventricular artery	}	Left atrium Left ventricle Interventricular septum
Right coronary artery	Marginal artery Posterior interventricular artery* AV nodal artery SA nodal artery	}	Right atrium Right ventricle Interventricular septum AV node SA node

* The heart blood supply is considered **right-side dominant** if the posterior interventricular artery arises from the right coronary artery. The heart blood supply is considered **left-side dominant** if the posterior interventricular artery arises from the left coronary artery.

Table 2-4.
Venous Drainage of the Heart

Vein	Travels With	Drains Into
Great cardiac vein	Anterior interventricular artery	Coronary sinus
Middle cardiac vein	Posterior interventricular artery	Coronary sinus
Small cardiac vein	Marginal artery	Coronary sinus
Anterior cardiac veins		Right atrium
Least cardiac veins		Nearest heart chamber

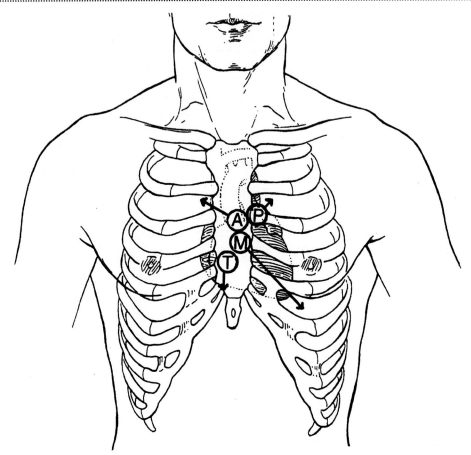

Figure 2-6. Position of the valves of the heart and heart sounds. P = pulmonary valve; A = aortic valve; M = mitral valve; T = tricuspid valve. Arrows indicate position of auscultation sites. Reproduced with permission from Chung, KW: *BRS Gross Anatomy*, 2nd edition, Baltimore, Williams & Wilkins, 1991.

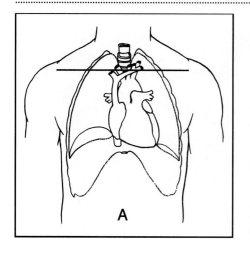

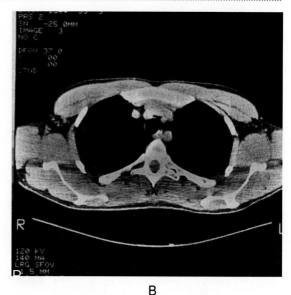

B

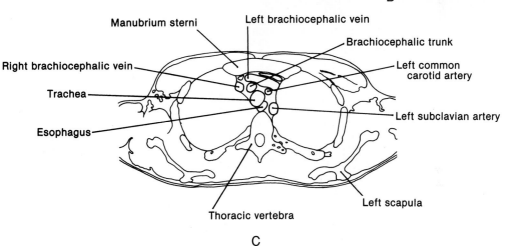

C

Figure 2-7.   (A) A schematic illustrating the T2-T3 level of the cross section. Reproduced with permission from Barrett, CP ; Poliakoff, SJ; Holder, LE; et al.: *Primer of Sectional Anatomy with MRI and CT Correlation,* 2nd edition, Baltimore, Williams & Wilkins, 1994. (B) A CT scan demonstrating the relationships of the brachiocephalic trunk, left common carotid artery, and left subclavian artery. (C) A schematic representation of the CT scan. Reproduced with permission from Slaby, F and Jacobs, ER: *Radiographic Anatomy,* Baltimore, Williams & Wilkins, 1990.

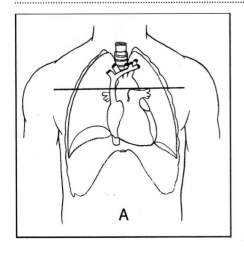

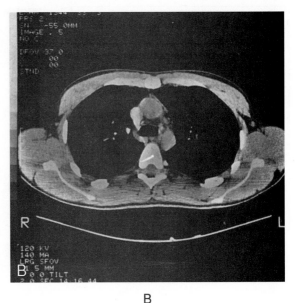

Superior
vena cava

Aortic arch
near its origin

Aortic–pulmonary
window

Aortic arch
near its
termination

Azygos vein
arching over
right primary bronchus

Trachea just above
the carina

Esophagus

C

Figure 2-8. (A) A schematic illustrating the T6 level of the cross section. Reproduced with permission from Barrett, CP ; Poliakoff, SJ; Holder, LE; et al.: *Primer of Sectional Anatomy with MRI and CT Correlation*, 2nd edition, Baltimore, Williams & Wilkins, 1994. (B) A CT scan demonstrating the aorticopulmonary window that extends from the bifurcation of the pulmonary trunk to the undersurface of the aortic arch. (C) A schematic representation of the CT scan. Reproduced with permission from Slaby, F and Jacobs, ER: *Radiographic Anatomy*, Baltimore, Williams & Wilkins, 1990.

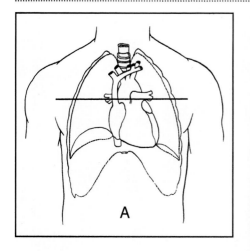

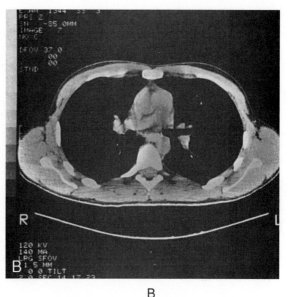

B

Pulmonary trunk

Ascending aorta

Level of division of left main stem
bronchus into upper and
lower lobar bronchi

Superior
vena cava

Right
pulmonary artery

Intermediate bronchus

Descending
thoracic aorta

Esophagus

C

Figure 2-9.   (A) A schematic illustrating the T7 level of the cross section. Reproduced with permission from Barrett, CP ; Poliakoff, SJ; Holder, LE; et al.: *Primer of Sectional Anatomy with MRI and CT Correlation,* 2nd edition, Baltimore, Williams & Wilkins, 1994. (B) A CT scan demonstrating right pulmonary artery, pulmonary trunk, and ascending aorta. (C) A schematic representation of the CT scan. Reproduced with permission from Slaby, F and Jacobs, ER: *Radiographic Anatomy,* Baltimore, Williams & Wilkins, 1990.

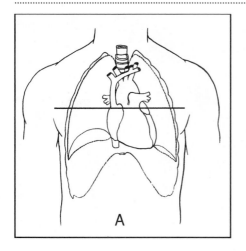

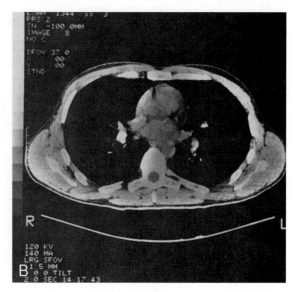

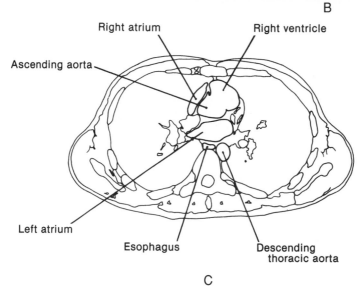

Figure 2-10. (A) A schematic illustrating the T8 level of the cross section. Reproduced with permission from Barrett, CP ; Poliakoff, SJ; Holder, LE; et al.: *Primer of Sectional Anatomy with MRI and CT Correlation,* 2nd edition, Baltimore, Williams & Wilkins, 1994. (B) A CT scan demonstrating the ascending aorta, right atrium, and right ventricle. (C) A schematic representation of the CT scan. Note the close proximity of the left atrium to the esophagus. Reproduced with permission from Slaby, F and Jacobs, ER: *Radiographic Anatomy,* Baltimore, Williams & Wilkins, 1990.

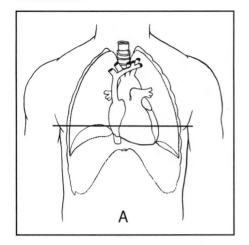

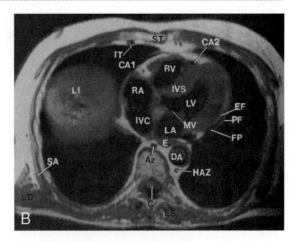

KEY

Az azygos vein
C thoracic spinal cord
CA1 branch of right coronary artery
CA2 branch of left anterior
   descending coronary artery
DA descending aorta
E esophagus
EF epicardial fat

ES erector spinae
FP fibrous pericardium
HAZ hemiazygos vein
IT internal thoracic vessels
IVC inferior vena cava
IVS interventricular septum
LA left atrium
LD latissimus dorsi
LI liver

LV left ventricle
MV mitral valve
PF pericardial fat
RA right atrium
RV right ventricle
SA serratus anterior
ST sternum

Figure 2-11.   (A) A schematic illustrating the T10 level of the cross section. (B) A T1 magnetic resonance image (MRI) demonstrating all four chambers of the heart. Note the close relationship of the left atrium and esophagus. Reproduced with permission from Barrett, CP ; Poliakoff, SJ; Holder, LE; et al.: *Primer of Sectional Anatomy with MRI and CT Correlation,* 2nd edition, Baltimore, Williams & Wilkins, 1994.

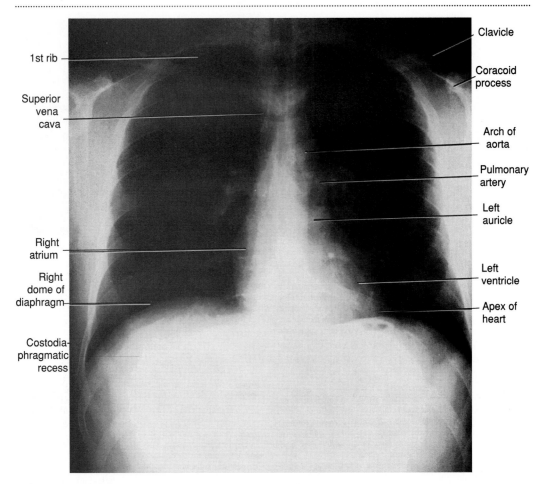

Figure 2-12.   A PA radiograph of the chest showing the structures that make up the right and left borders of the heart. Note that the dome of the diaphragm is somewhat higher on the right. Reproduced with permission from Moore, KL: *Clinically Oriented Anatomy*, 3rd edition, Baltimore, Williams & Wilkins, 1992.

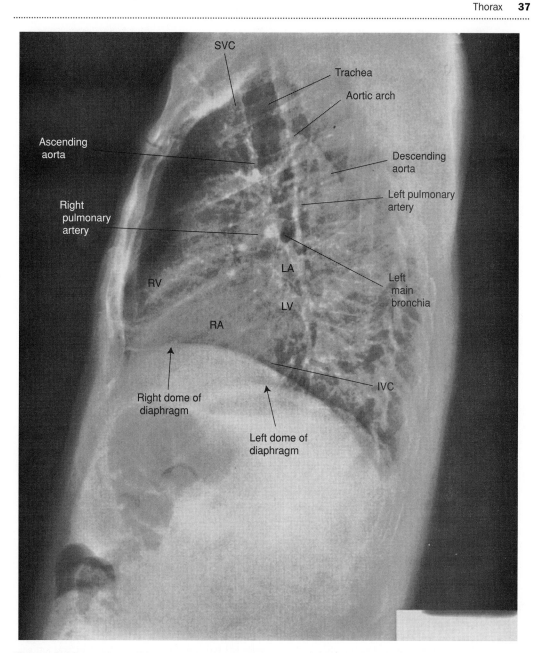

Figure 2-13. A lateral chest radiograph. A lateral chest x-ray generally is done with the patient's left side against the film. The cardiac silhouette on a lateral chest radiograph consists of the left atrium (*LA*; upper posterior edge), left ventricle (*LV*; lower posterior edge), and right ventricle (*RV*; anterior edge). The right atrium (*RA*) forms no edge on a lateral projection but is seen en face. The left main bronchus can be observed with the left pulmonary artery arching over it and the right pulmonary artery lying anterior. The area between the undersurface of the aortic arch and the left pulmonary artery as seen on a lateral projection is called the aorticopulmonary window. Reproduced with permission from Slaby, F and Jacobs, ER: *Radiographic Anatomy*, Baltimore, Williams & Wilkins, 1990.

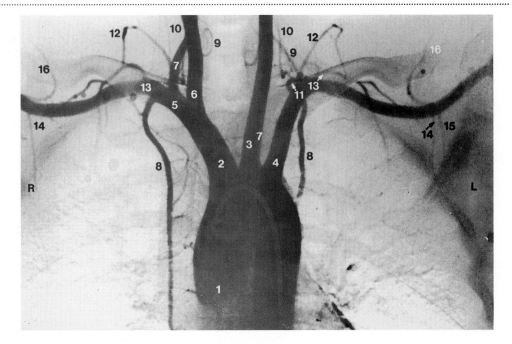

1 Ascending aorta
2 Brachiocephalic trunk
3 Left common carotid artery
4 Left subclavian artery
5 Right subclavian artery
6 Right common carotid artery
7 Vertebral artery
8 Internal thoracic artery
9 Inferior thyroid artery
10 Ascending cervical artery
11 Thyrocervical trunk
12 Suprascapular artery
13 Costocervical trunk
14 Superior thoracic artery
15 Lateral thoracic artery
16 Deltoid branch of thoraco-

Figure 2-14.   Aortic angiogram. Observe the ascending aorta (*1*), aortic arch, and descending aorta. Note the three main branches of the arch of the aorta, the brachiocephalic trunk (*2*), left common carotid artery (*3*), and left subclavian artery (*4*). The brachiocephalic trunk branches into the right subclavian (*5*) and right common carotid (*6*) arteries. The right and left vertebral arteries (*7*) typically branch from the right and left subclavian arteries, respectively. In this patient, however, the left vertebral artery branches directly from the arch of the aorta (a normal variant). If contrast material is injected into the **right** subclavian artery, the entire circle of Willis is demonstrated because contrast material travels up both the right vertebral artery and right internal carotid artery. If contrast material is injected into the **left** subclavian artery, typically only the posterior portion of the circle of Willis is demonstrated because contrast material travels up only the left vertebral artery. In this particular patient, no part of the circle of Willis is demonstrated if contrast material is injected into the left subclavian artery due to the variant left vertebral artery arising from the arch of the aorta. Reproduced with permission from Weir, J and Abrahams, PH: *An Imaging Atlas of Human Anatomy*, London, UK, Mosby International, 1992.

1 Left main stem coronary artery
2 Left anterior interventricular
  branch (left anterior
  descending)
3 Circumflex artery

4 First   ⎫ obtuse marginal
5 Second ⎬ branch of
         ⎭ circumflex artery

7 Diagonal arteries
9 Septal arteries

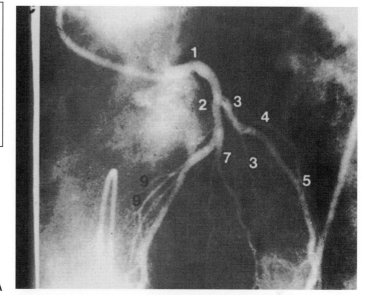

A

1 Right coronary artery
2 Conus artery
3 Sinuatrial nodal artery
4 Right marginal arteries
5 Posterior interventricular septal
  artery (posterior descending
  artery)
6 Atrioventricular nodal artery
7 Lateral ventricular branch to
  left ventricle

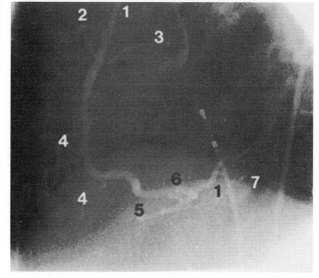

B

Figure 2-15.   (A) Arteriogram of the left coronary artery and its branches from a left anterior oblique view. (B) Arteriogram of the right coronary artery and its branches from a lateral view. Reproduced with permission from Weir, J and Abrahams, PH: *An Imaging Atlas of Human Anatomy*, London, UK, Mosby International, 1992.

3

Abdomen

I. ABDOMINAL WALL

A. Abdominal regions

1. The abdomen can be topographically divided into **nine regions:**

   a. Right hypochondriac

   b. Epigastric

   c. Left hypochondriac

   d. Right lumbar

   e. Umbilical

   f. Left lumbar

   g. Right inguinal

   h. Hypogastric

   i. Left inguinal

2. In order to decipher many clinical vignette questions, it is necessary to know which organs are associated with each region (Figure 3-1).

B. Penetration of the abdomen. Paracentesis (Figure 3-2) is a procedure that uses a needle inserted through various layers of the abdominal wall to withdraw excess peritoneal fluid. **Knife wounds** to the abdomen penetrate these same layers. There are two approaches:

1. **Midline approach.** The needle or knife passes through the following structures:

   a. Skin

   b. Superficial fascia (Camper's and Scarpa's)

   c. Linea alba

   d. Transversalis fascia

   e. Extraperitoneal fat

   f. Parietal peritoneum

2. **Flank approach.** The needle or knife passes through the following structures:

   a. Skin

   b. Superficial fascia (Camper's and Scarpa's)

   c. External oblique muscle

   d. Internal oblique muscle

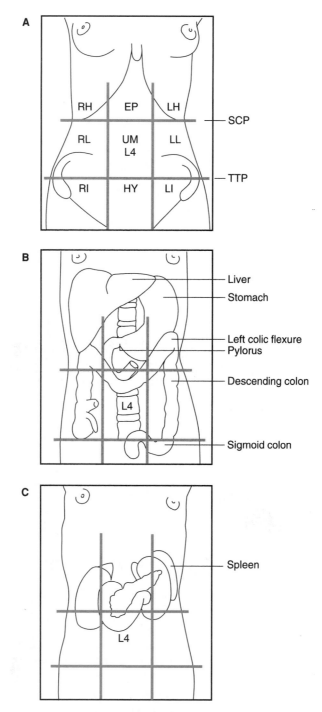

Figure 3-1. (A) A commonly used clinical method for subdividing the abdomen into specific regions using the subcostal plane (SCP), transtubercular plane (TTP; joining the tubercle of the iliac crests), and the mid-clavicular lines (MC). RH: right hypochondriac; EP: epigastric; LH: left hypochondriac; RL: right lumbar; UM: umbilical; LL: left lumbar; RI: right inguinal; HY: hypogastric; LI: left inguinal. (B) Surface projection of the stomach, liver, and large intestine. Outlines of vertebrae are shown for orientation. (C) Surface projection of the duodenum, pancreas, kidneys, suprarenal gland, and spleen. Note that the U-shaped duodenum lies superior to the umbilicus and that the two ends are not far apart. Adapted with permission from Moore, KL: *Clinically Oriented Anatomy*, 3rd edition, Baltimore, Williams & Wilkins, 1992.

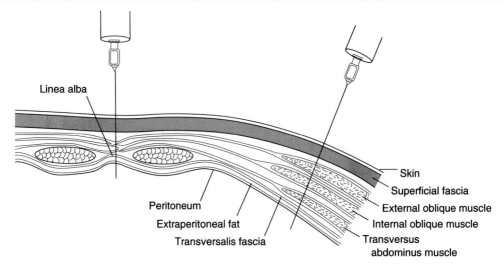

Linea alba

Peritoneum

Extraperitoneal fat

Transversalis fascia

Skin

Superficial fascia

External oblique muscle

Internal oblique muscle

Transversus abdominus muscle

Figure 3-2. A transverse section through the anterior abdominal wall demonstrating the various layers that would be penetrated by paracentesis or a knife wound. Adapted with permission from Moore, KL: *Clinically Oriented Anatomy*, 3rd edition, Baltimore, Williams & Wilkins, 1992.

    e. Transverse abdominus muscle

    f. Transversalis fascia

    g. Extraperitoneal fat

    h. Parietal peritoneum

C. Inguinal region is an area of weakness of the anterior abdominal wall resulting from the penetration of the testes and spermatic cord (in males) or the round ligament of the uterus (in females) during embryological development.

    1. Inguinal ligament is the coiled lower border of the aponeurosis of the **external oblique muscle** and extends from the anterior–superior iliac spine to the pubic tubercle.

    2. Deep inguinal ring is an oval opening in the **transversalis fascia** located lateral to the inferior epigastric artery.

    3. Superficial inguinal ring is a triangular defect in the aponeurosis of the **external oblique muscle** located lateral to the pubic tubercle.

    4. Inguinal canal is a canal that begins at the deep inguinal ring and ends at the superficial inguinal ring. This canal transmits the spermatic cord (in males) or round ligament of the uterus (in females). The main components of the walls of the inguinal canal are shown in Table 3-1.

    5. Hernias may occur in the inguinal region or other areas of weakness. The characteristics of various types of hernias are shown in Table 3-2.

D. Scrotum (Figure 3-3) is an outpouching in the lower abdominal wall whereby layers of the abdominal wall continue into the scrotal area to cover the spermatic cord and testes.

    1. Cancer of the scrotum metastasizes to **superficial inguinal nodes.**

    2. Cancer of the testes metastasizes to **deep lumbar nodes** resulting from the embryological development of the testes within the abdominal cavity and subsequent descent into the scrotum.

Table 3-1.

Main Components of Inguinal Canal Walls

Wall of the Inguinal Canal	Main Components
Anterior	Aponeurosis of the external oblique muscle Aponeurosis of the internal oblique muscle
Posterior	Transversalis fascia Conjoint tendon (aponeuroses of the internal oblique and transverse abdominus muscles)
Roof	Internal oblique muscle Transverse abdominus muscle
Floor	Inguinal ligament (aponeurosis of the external oblique muscle) Lacunar ligament (medial expansion of inguinal ligament)

II. THE PERITONEAL CAVITY is divided into the lesser peritoneal sac (omental bursa) and greater peritoneal sac.

   A. The lesser peritoneal sac forms due to the 90-degree clockwise rotation of the stomach during embryological development. The lesser sac boundaries include:

   1. Anteriorly—the liver, stomach, and lesser omentum

   2. Posteriorly—the diaphragm

   3. Right side—liver

   4. Left side—gastrosplenic and lienorenal ligaments

   B. The **greater peritoneal sac** is the remainder of the peritoneal cavity and extends from the diaphragm to the pelvis. The greater sac contains a number of pouches, recesses, and paracolic gutters through which peritoneal fluid circulates.

Table 3-2.

Hernia Characteristics

Type of Hernia	Characteristics
Direct inguinal hernia	Protrudes directly throught the anterior abdominal wall within Hesselbach's triangle* Protrudes *medial* to the inferior epigastric artery and vein** Common in older men; rare in women
Indirect inguinal hernia	Protrudes through the deep inguinal ring to enter the inguinal canal and may exit through the surperficial inguinal ring Protrudes *lateral* to the inferior epigastric artery and vein** Protrudes *above and medial* to the pubic turbercle*** Common in males, young adults, children More common than a direct inguinal hernia
Femoral hernia	Protrudes through the femoral canal below the inguinal ligament Protrudes *below and lateral* to the pubic tubercle*** More common in females

* Hesselbach's (inguinal) triangle is bound laterally by the inferior epigastric artery and vein, medially by the rectus abdominus muscle, and inferiorly by the inguinal ligament.
** Distinguishing feature of a direct hernia versus an indirect hernia
*** Distinguishing feature of an indirect hernia versus a femoral hernia

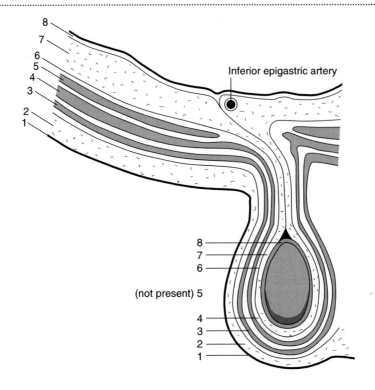

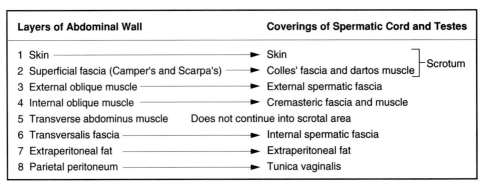

Layers of Abdominal Wall	Coverings of Spermatic Cord and Testes
1 Skin ⟶	Skin
2 Superficial fascia (Camper's and Scarpa's) ⟶	Colles' fascia and dartos muscle
3 External oblique muscle ⟶	External spermatic fascia
4 Internal oblique muscle ⟶	Cremasteric fascia and muscle
5 Transverse abdominus muscle	Does not continue into scrotal area
6 Transversalis fascia ⟶	Internal spermatic fascia
7 Extraperitoneal fat ⟶	Extraperitoneal fat
8 Parietal peritoneum ⟶	Tunica vaginalis

(Skin and Colles' fascia and dartos muscle are bracketed as Scrotum)

Figure 3-3. A schematic horizontal section showing the inguinal canal and the derivation of the coverings of the spermatic cord and testes. Adapted with permission from Moore, KL: *Clinically Oriented Anatomy*, 3rd edition, Baltimore, Williams & Wilkins, 1992.

1. **Paracolic gutters** are channels that run along the ascending and descending colon. Normally, peritoneal fluid flows **upward** through the paracolic gutters to the **subphrenic recess,** where it enters the lymphatics associated with the diaphragm.

2. **Excess peritoneal fluid** due to peritonitis or ascites flows:

   a. **Downward** through the paracolic gutters to the **rectovesical pouch** (in males) or the **rectouterine pouch** (in females) when the patient is in a **sitting** or **standing** position

   b. Upward through the paracolic gutters to the **subphrenic recess** and the **hepatorenal recess** when the patient is in the **supine** position

   (1) The patient may complain of **shoulder pain** (referred pain) because the supraclavicular nerves and the phrenic nerves have the same nerve root origins (C3, C4, C5).

   (2) The hepatorenal recess is the **lowest** part of the peritoneal cavity when the patient is in the supine position.

C. Epiploic (Winslow's) foramen is the opening (or connection) between the lesser peritoneal sac and greater peritoneal sac. If a surgeon places his finger in the epiploic foramen, the **inferior vena cava (IVC)** will lie posterior and the **portal vein** will lie anterior.

III. OMENTUM (Figure 3-4)

A. Lesser omentum

   1. The lesser omentum extends from the **porta hepatis** of the liver to the **lesser curvature** of the stomach.

   2. The lesser omentum consists of the **hepatoduodenal ligament** and **hepatogastric ligament.**

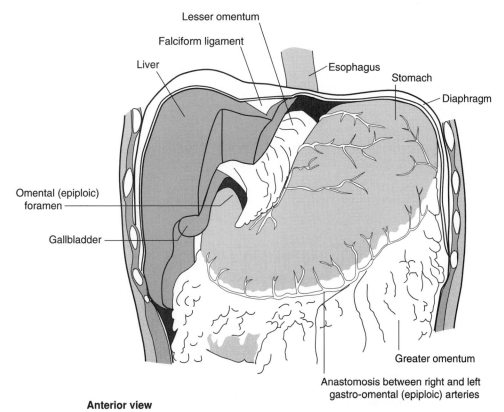

Figure 3-4.   An anterior dissection of the stomach and the omenta associated with it. The left part of the liver is cut away to expose the epiploic (Winslow's) foramen, lesser omentum, and the portal triad. Adapted with permission from Moore, KL: *Clinically Oriented Anatomy*, 3rd edition, Baltimore, Williams & Wilkins, 1992.

3. The **portal triad** lies in the free margin of the hepatoduodenal ligament and consists of the:

    a. **Portal vein**—posterior

    b. **Common bile duct**—anterior and to the right

    c. **Hepatic artery**—anterior and to the left

B. **Greater omentum** hangs down from the greater curvature of the stomach.

IV. INTRAPERITONEAL AND EXTRAPERITONEAL VISCERA (Table 3-3)

V. ABDDOMINAL AORTA (Figure 3-5)

A. **Branches** (Table 3-4)

B. **Gradual occlusion** (due to atherosclerosis) at the bifurcation of the abdominal aorta (a common site) may result in **claudication** (pain in the legs when walking) and **impotence** (due to the lack of blood to the internal iliac arteries).

C. **Aneurysms** most commonly occur between the renal arteries and the bifurcation of the abdominal aorta.

D. **Routes of collateral circulation** exist in case the abdominal aorta is blocked.

    1. Internal thoracic artery → superior epigastric artery → inferior epigastric artery

    2. Superior pancreaticoduodenal artery [from celiac trunk (CT)] → inferior pancreatico-duodenal artery [from superior mesenteric artery (SMA)]

    3. Middle colic artery (from SMA) → left colic artery [from inferior mesenteric artery (IMA)]

    4. Marginal artery (from SMA and IMA)

    5. Superior rectal artery (from IMA) → middle rectal artery (from internal iliac artery)

Table 3-3.

Intraperitoneal and Extraperitoneal Viscera

Intraperitoneal	Extraperitoneal
Stomach	Part 2, 3, 4 of duodenum
Part 1 of duodenum	Ascending colon
Jejunum	Descending colon
Ileum	Rectum
Cecum	Head, neck, body of pancreas
Appendix	Kidneys
Transverse colon	Ureters
Sigmoid colon	Suprarenal gland
Liver	Abdominal aorta
Gallbladder	Inferior vena cava
Tail of pancreas	
Spleen	

VISCERAL BRANCHES

PARIETAL BRANCHES

Paired   **Unpaired**

Paired   **Unpaired**

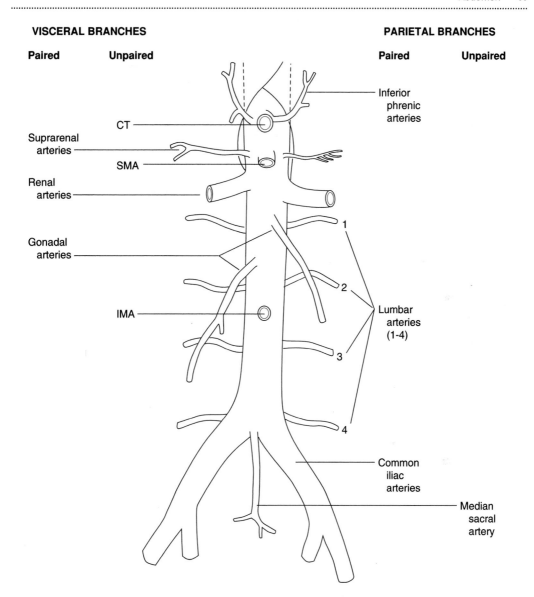

CT

Suprarenal
arteries

SMA

Renal
arteries

Gonadal
arteries

IMA

Inferior
phrenic
arteries

1

2

Lumbar
arteries
(1-4)

3

4

Common
iliac
arteries

Median
sacral
artery

Figure 3-5.   An anterior view of the abdominal aorta and its branches. CT: celiac trunk; SMA: superior mesenteric artery; IMA: inferior mesenteric artery. Adapted with permission from Moore, KL: *Clinically Oriented Anatomy*, 3rd edition, Baltimore, Williams & Wilkins, 1992.

Table 3-4.

Branches of the Abdominal Aorta

Artery	Type	Structures Supplied
Suprarenal arteries	Visceral, paired	Suprarenal gland
Renal arteries	Visceral, paired	Kidney
Gonadal arteries	Visceral, paired	Testes, ovaries

Celiac trunk (CT)
  Left gastric artery
  Splenic artery
  Common hepatic artery
	Visceral, unpaired	Foregut Derivatives:
		  Esophagus
		  Stomach
		  Duodenum
		  Liver and gallbladder
		  Pancreas
		Spleen*

Superior mesenteric artery (SMA)	Visceral, unpaired	Midgut derivatives:
		  Duodenum
		  Jejunum
		  Ileum
		  Cecum
		  Appendix
		  Ascending colon
		  Proximal $2/3$ of transverse colon

Inferior mesenteric artery (IMA)	Visceral, unpaired	Hindgut derivatives:
		  Distal $1/3$ of transverse colon
		  Descending colon
		  Sigmoid colon
		  Upper portion of rectum

Inferior phrenic arteries	Parietal, paired	Diaphragm
Lumbar arteries**	Parietal, paired	Body wall
Common iliac arteries	Parietal, paired	Pelvis, perineum, leg
Median sacral artery	Parietal, unpaired	Body wall

* The spleen is *not* a foregut derivative, but is supplied by the celiac trunk.
** May give off the great radicular artery (Important blood supply to the spinal cord)

VI. VENOUS DRAINAGE OF ABDOMEN

A. Azygos venous system

1. The **azygos vein** ascends on the right side of the vertebral column and drains blood from the IVC to the superior vena cava (SVC).

2. The **hemiazygos vein** ascends on the left side of the vertebral column and drains blood from the left renal vein to the azygous vein.

B. **Inferior vena cava (IVC)** [Figure 3-6]

1. The IVC is **formed** by the union of the right and left common iliac veins at vertebral level L5.

2. The IVC **drains all the blood** from below the diaphragm (even portal blood from the gastrointestinal tract after it percolates through the liver) to the right atrium.

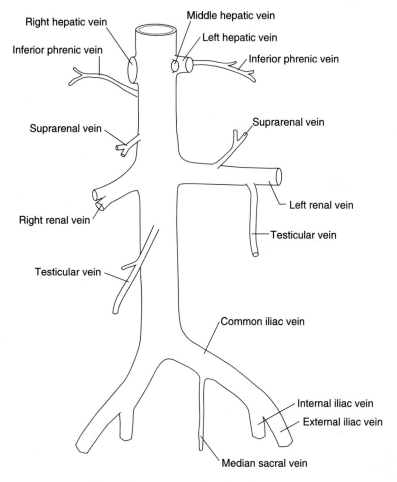

Figure 3-6.   An anterior view of the inferior vena cava (IVC) and its tributaries. Adapted with permission from Moore, KL: *Clinically Oriented Anatomy,* 3rd edition, Baltimore, Williams & Wilkins, 1992.

3. The IVC is **in jeopardy during surgical repair** of a herniated intervertebral disc.

4. The **right gonadal** and **right suprarenal veins** drain directly into the IVC, whereas the **left gonadal** and **left suprarenal veins** drain into the left renal vein.

5. The appearance of a **left testicular varicocele** in the male may indicate occlusion of the **left testicular vein** and/or **left renal vein** (caused by a malignant tumor of the kidney, for example).

6. **Routes of collateral circulation** exist in case the IVC is blocked [e.g., malignant retroperitoneal tumors, large blood clots (thrombi)]. These routes include:

    a. Femoral vein → superficial epigastric vein → lateral thoracic vein → axillary vein → right atrium

    b. Hemiazygos vein → azygos vein → superior vena cava → right atrium

    c. Lumbar veins → vertebral venous plexuses → cranial dural sinuses → internal jugular vein → right atrium

  d. External iliac vein → superior epigastric vein → internal thoracic vein → brachiocephalic vein → right atrium

 C. **Hepatic portal system** (Figure 3-7)

  1. In general, the term *portal* refers to a vein interposed between two capillary beds (e.g., capillary bed → vein → capillary bed)

  2. The hepatic portal system consists specifically of the following **vascular structures:**

   a. Capillary bed of the gastrointestinal tract

   b. Portal vein

   c. Capillary bed of the liver (hepatic sinusoids)

  3. The **portal vein** is formed **posterior to the neck of the pancreas** by the union of the **splenic vein** and **superior mesenteric vein.** The **inferior mesenteric vein** usually ends by joining the splenic vein.

  4. The **blood** within the portal vein carries high levels of **nutrients** from the gastrointestinal tract and **products** of red blood cell destruction from the spleen.

 D. **Portal–IVC (caval) anastomosis** (see Figure 3-7) becomes clinically relevant when **portal hypertension** (resulting from liver cirrhosis, for example) occurs. Portal hypertension causes blood within the portal vein to reverse its flow and enter the IVC in order to return to the heart. There are **three main sites** of portal–IVC anastomosis (Table 3-5).

VII. ESOPHAGUS

 A. **Tumors, foreign bodies,** and **chemical burns** have a predisposition to localize at three anatomical sites where the esophagus is naturally constricted. These sites include:

  1. Junction between the pharynx and esophagus

  2. Level of tracheal bifurcation

  3. Junction between the esophagus and stomach

 B. The esophagus may also be constricted due to an **enlarged left atrium.**

 C. **Malignant tumors** of the lower one third of the esophagus tend to spread below the diaphragm to the celiac lymph nodes.

 D. A **sliding hiatal hernia** occurs when the esophagus and stomach are pulled up through the diaphragm.

 E. A **paraesophageal hiatal hernia** occurs when a portion of the stomach is pulled up through the diaphragm.

Table 3-5.

Sites of Portal–IVC Anastomosis

Site of Anastomosis	Clinical Sign	Veins Involved in Portal–IVC Anastomosis
Esophagus	Esophageal varices	Left gastric vein ↔ esophageal vein
Umbilicus	Caput medusa	Paraumbilical vein ↔ superficial and inferior epigastric veins
Rectum	Hemorrhoids	Superior rectal vein ↔ middle and inferior rectal veins

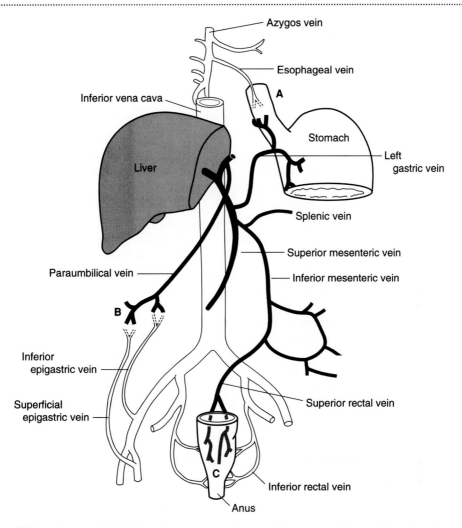

Figure 3-7. The portal–IVC (caval) anastomoses. In case of liver or portal vein obstruction, these anastomoses provide collateral circulation through the IVC back to the heart. A, B, C are the three main sites of anastomoses. Adapted with permission from Moore, KL: *Clinically Oriented Anatomy*, 3rd edition, Baltimore, Williams & Wilkins, 1992.

VIII. STOMACH

A. G cells secrete **gastrin,** which stimulates HCl secretion from parietal cells. G cells are found predominately in the **pyloric antrum** of the stomach.

B. The **parasympathetic innervation** of the stomach is via branches of the vagus nerve (X), most notably the **anterior** and **posterior nerves of Latarget,** which run along the **lesser curvature** of the stomach. HCl secretion from parietal cells is under parasympathetic control.

C. Gastric ulcers may be treated by:

1. Surgical removal of the pyloric antrum, which eliminates the hormonal stimulation of HCl secretion

   2. Selective vagotomy of the nerves of Latarget, which eliminates the neural stimulation of HCl secretion

   D. The **"dumping syndrome"** refers to an abnormally rapid emptying of the stomach contents, usually following partial gastrectomy or vagotomy.

IX. DUODENUM. The duodenum is divided into four parts.

   A. First (superior) part

   1. The superior part **begins** at the pylorus of the stomach (**gastroduodenal junction**), which is marked by the **prepyloric vein** (an important clinical landmark). This part has a mesentery and, therefore, is mobile.

   2. Posterior relations include the **common bile duct** and **gastroduodenal artery.**

   3. Radiologists refer to the beginning of this part of the duodenum as the **duodenal cap** or **bulb. Duodenal ulcers** commonly occur at the duodenal cap and may perforate the posterior wall and erode the gastroduodenal artery, causing severe hemorrhage.

   B. The **second (descending)** part is retroperitoneal and receives the **common bile duct** and **main pancreatic duct** on its posterior/medial wall at the **hepatopancreatic ampulla** (Vater's ampulla).

   C. The **third (horizontal)** part is retroperitoneal and runs horizontally across the L3 vertebra between the superior mesenteric artery anteriorly and the aorta and IVC posteriorly. In **severe abdominal injuries,** this part of the duodenum may be crushed against the L3 vertebra.

   D. The **fourth (ascending)** part ascends to meet the jejunum at the duodenojejunal flexure, which is supported by the **suspensory ligament of Treitz.** Treitz's ligament represents the cranial end of the dorsal mesentery.

X. FEATURES OF JEJUNUM, ILEUM, AND LARGE INTESTINE (Table 3-6)

XI. BILIARY DUCTS (Figure 3-8)

   A. The **cystic duct** drains bile from the gallbladder. The **mucosa** of the cystic duct is arranged in a spiral fold with a core of smooth muscle known as the **spiral valve.** The spiral valve keeps the cystic duct **constantly open** so that bile can flow freely in either direction.

   B. There are **three clinically important sites** of **gallstone obstruction.**

   1. Within the cystic duct

   a. A stone may transiently lodge within the cystic duct and cause pain (**biliary colic**) within the **epigastric region** due to the distention of the duct.

   b. If a stone becomes entrapped within the cystic duct, bile flow from the gallbladder will be obstructed, resulting in inflammation of the gallbladder (**acute cholecystitis**), and pain will shift to the **right hypochondriac region.** Bile flow from the liver remains open.

   2. Within the common bile duct. If a stone becomes entrapped within the common bile duct, bile flow from both the gallbladder and liver will be obstructed, resulting in inflammation of the gallbladder and liver. **Jaundice** frequently is observed.

Table 3-6.
Features of the Jejunum, Ileum, and Large Intestine

Jejunum	Ileum	Large Intestine
Villi present	Villi present	Villi absent
Intestinal glands (crypts) present	Intestinal glands (crypts) present	Intestinal glands (crypts) absent
Initial $^2/_5$ of small intestine	Terminal $^3/_5$ of small intestine	**Tenia coli** (3 longitudinal bands of smooth muscle) present
Resides in umbilical region	Resides in hypogastric and inguinal regions	**Appendices epiploicae** (fatty tags) present
Larger diameter than ileum	Smaller diameter than jejunum	**Haustra** (sacculations of the wall) present
Vasa recta are longer than those to ileum	Vasa recta are shorter than those to jejunem	
Large plica circularis* (can be palpated)	Small plica circularis	
Often empty		
Thicker, more vascular, and redder in the living person than ileum		

* Plica circularis are folds of the mucosa and submucosa.

3. At the hepatoduodenal ampulla. If a stone becomes entrapped at the ampulla, bile flow from both the gallbladder and liver will be obstructed. In addition, bile may pass into the pancreatic duct. **Jaundice** and **pancreatitis** frequently are observed.

XII. LIVER

A. The liver is divided into the **right lobe** and **left lobe** by the **falciform ligament.**

B. Structures within the portal triad enter and leave the liver between the **caudate** and **quadrate lobes.**

C. The liver is **secured** in its anatomical location by the **attachment** of the hepatic veins to the IVC, which allows for very little rotation of the liver during surgery.

D. Liver biopsies frequently are performed by needle puncture through the right intercostal space 8, 9, or 10. The needle passes through the following structures:

1. Skin

2. Superficial fascia

3. External oblique muscle

4. Intercostal muscles

5. Costal parietal pleura

6. Costodiaphragmatic recess

7. Diaphragmatic parietal pleura

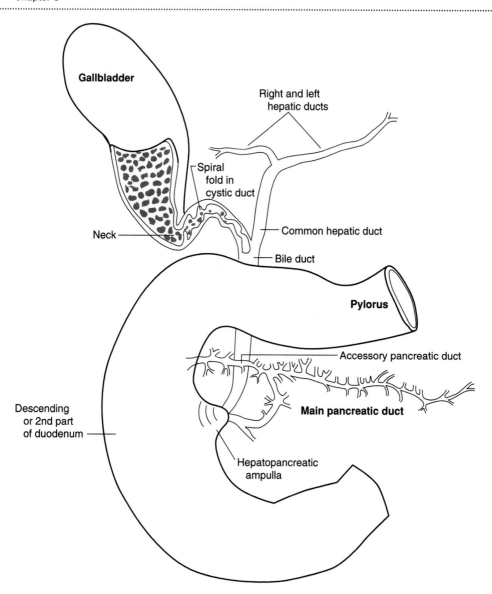

Figure 3-8.   Anterior view of a dissection of the various extrahepatic bile ducts and pancreatic ducts. Note the cystic duct with its spiral fold (or spiral valve). Adapted with permission from Moore, KL: *Clinically Oriented Anatomy*, 3rd edition, Baltimore, Williams & Wilkins, 1992.

8. Diaphragm

9. Peritoneum

XIII. CROSS-SECTIONAL ANATOMY

A. At about T12 where the portal triad is located (Figure 3-9)

B. At the level of the gallbladder (Figure 3-10)

C. At the level of the hilum of the kidneys (Figure 3-11)

XIV. RADIOLOGY

A. Radiograph of the stomach and duodenum after a barium meal (Figure 3-12)

B. AP radiograph of the abdomen after a barium enema (Figure 3-13)

C. Arteriogram of the celiac trunk (Figure 3-14)

D. Arteriogram of the superior mesenteric artery (Figure 3-15)

E. Pyelogram of the kidney (Figure 3-16)

F. Arteriogram of the kidney (Figure 3-17)

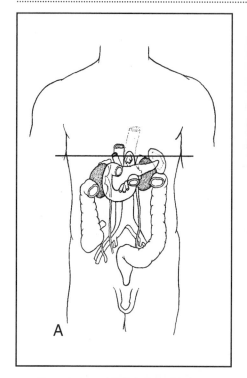

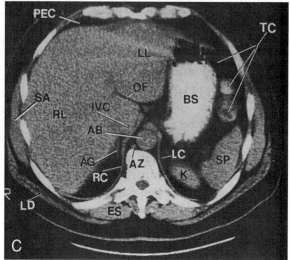

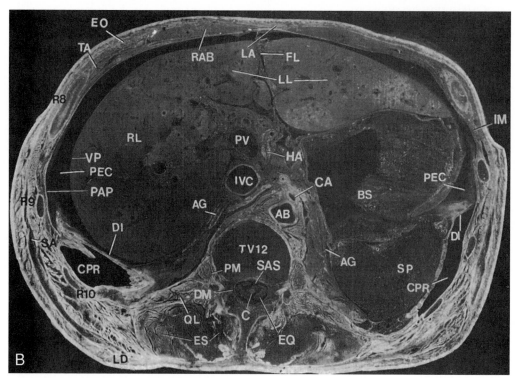

Figure 3-9.   *(continued)*

KEY
AB abdominal aorta
AG adrenal gland
AZ azygos vein
BS body of stomach
C sacral spinal cord
CA celiac artery
CPR posterior costophrenic recess
DI diaphragm
DM dura mater
EO external abdominal oblique
EQ cauda equina
ES erector spinae
FL falciform ligament

HA common hepatic artery
IM intercostal muscles
IVC inferior vena cava
K kidney
LA linea alba
LC left crus of diaphragm
LD latissimus dorsi
LL left lobe of liver
OF oblique fissure
PAP parietal peritoneum
PEC peritoneal cavity
PM psoas major
PV portal vein
QL quadratus lumborum

R8—10 ribs 8–10
RAB rectus abdominis
RC right crus of diaphragm
RL right lobe of liver
SA serratus anterior
SAS subarachnoid space (lumbar cistern)
SP spleen
TA transversus abdominis
TC transverse colon
TV12 body, thoracic vertebra 12
VP visceral peritoneum

Figure 3-9. A cross section and CT scan at about T12 where the portal triad is located. (A) A schematic to show where the cross section was taken. (B) A cross section through a cadaver. (C) A CT scan. Note the various structures as indicated by the key. In addition, note the psoas major and quadratus lumborum muscles along the sides of the vertebral body. The right and left lobes of the liver are shown in their relationship to the portal vein, common hepatic artery, and IVC. The right adrenal gland lies posterolateral to the IVC. The left adrenal gland lies between the body of the stomach and the abdominal aorta. Reproduced with permission from Barrett, CP ; Poliakoff, SJ; Holder, LE; et al.: *Primer of Sectional Anatomy with MRI and CT Correlation*, 2nd edition, Baltimore, Williams & Wilkins, 1994.

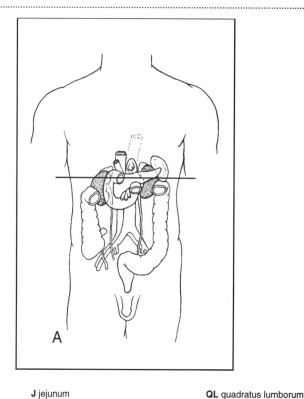

A

KEY

AB abdominal aorta
AG adrenal gland
BS body of stomach
D1 first part of duodenum
D2 second part of duodenum
D4 fourth part of duodenum
DC descending colon
EO external abdominal oblique
EQ cauda equina
ES erector spinae
FL falciform ligament
FT fat
GB gallbladder
IO internal abdominal oblique
IVC inferior vena cava

J jejunum
K kidney
LA linea alba
LC left crus of diaphragm
LD latissimus dorsi
LL left lobe of liver
LV2 body, lumbar vertebra 2
P1 head of pancreas
P2 body of pancreas
P3 tail of pancreas
PA antrum of stomach
PEC peritoneal cavity
PF perirenal fat
PM psoas major
PV portal vein
PY pyloric sphincter

QL quadratus lumborum
R10 rib 10
RA renal artery
RAB rectus abdominis
RC right crus of diaphragm
RL right lobe of liver
RV renal vein
SMA superior mesenteric artery
SMV superior mesenteric vein
SP spleen
SV splenic vein
TA transversus abdominis
TC transverse colon
TZ ligament of Treitz
UP uncinate process of pancreas

Figure 3-10.   A cross section and two CT scans at the level of the gallbladder. (A) A schematic to show where the cross section was taken. (B) A cross section through a cadaver. (C) (D) CT scans. Note the various structures as indicated by the key. The second part of the duodenum is adjacent to the head of the pancreas. The body of the pancreas extends to the left posterior to the stomach. The tail of the pancreas reaches the spleen. The uncinate process of the pancreas lies posterior to the superior mesenteric artery. The gallbladder lies between the right and left lobes of the liver just to the right of the antrum of the stomach. Reproduced with permission from Barrett, CP ; Poliakoff, SJ; Holder, LE; et al.: *Primer of Sectional Anatomy with MRI and CT Correlation*, 2nd edition, Baltimore, Williams & Wilkins, 1994.

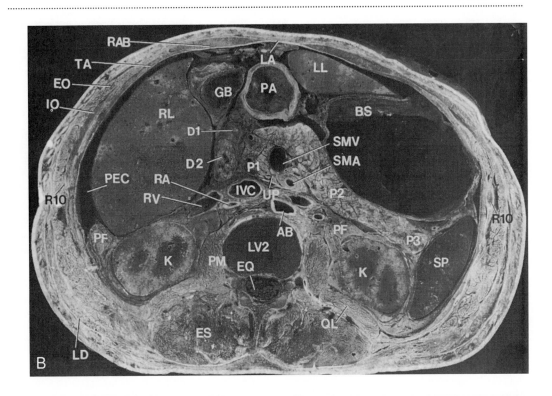

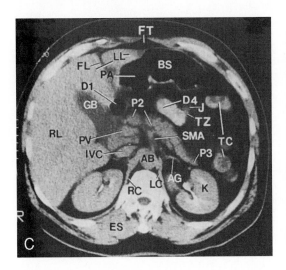

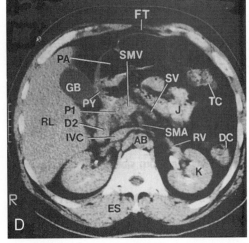

Figure 3-10. *(continued)*

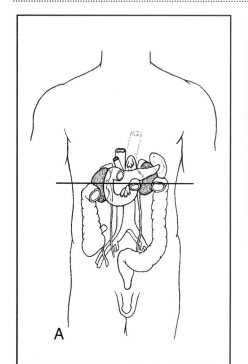

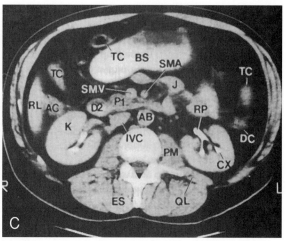

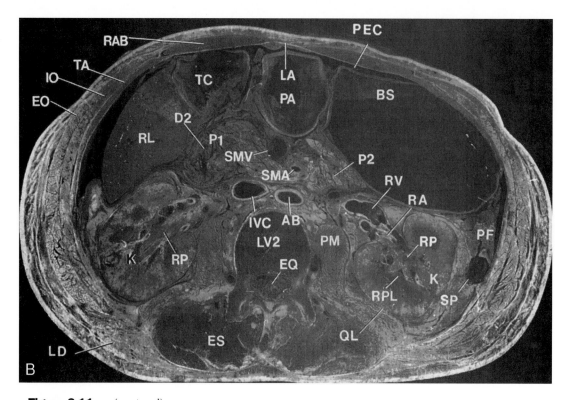

Figure 3-11. *(continued)*

KEY
AB abdominal aorta
AC ascending colon
BS body of stomach
CX renal calyx (minor)
D2 second part of duodenum
DC descending colon
EO external abdominal oblique
EQ cauda equina
ES erector spinae
IO internal abdominal oblique
IVC inferior vena cava

J jejunum
K kidney
LA linea alba
LD latissimus dorsi
LV2 body of LV2
P1 head of pancreas
P2 body of pancreas
PA antrum of stomach
PEC peritoneal cavity
PF perirenal fat
PM psoas major
QL quadratus lumborum

RA renal artery
RAB rectus abdominis
RL right lobe of liver
RP renal pelvis
RPL renal papilla
RV renal vein
SMA superior mesenteric artery
SMV superior mesenteric vein
SP spleen (lower tip)
TA transversus abdominis
TC transverse colon

Figure 3-11. A cross section and CT scan at the level of the hilum of the kidney. (A) A schematic to show where the cross section was taken. (B) A cross section through a cadaver. (C) A CT scan. Note the various structures as indicated by the key. The IVC and the abdominal aorta lie side by side as both vessels pass posterior to the pancreas. The second part of the duodenum contacts the right kidney and the right lobe of the liver. The left renal vein lies anterior to the renal artery. Reproduced with permission from Barrett, CP ; Poliakoff, SJ; Holder, LE; et al.: *Primer of Sectional Anatomy with MRI and CT Correlation*, 2nd edition, Baltimore, Williams & Wilkins, 1994.

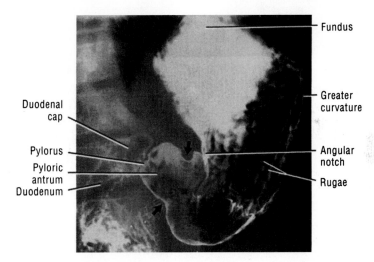

Figure 3-12. A radiograph of the stomach and duodenum after a barium meal. Note the structures indicated. The bold arrows demonstrate a peristaltic wave. Reproduced with permission from Moore, KL: *Clinically Oriented Anatomy*, 3rd edition, Baltimore, Williams & Wilkins, 1992.

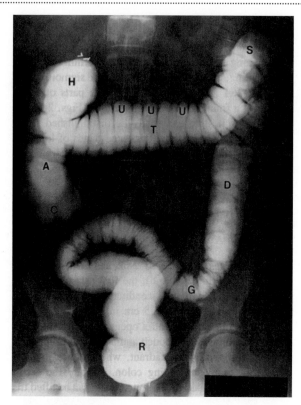

Figure 3-13.   A radiograph of the abdomen following a barium enema. *C* = cecum; *A* = ascending colon; *H* = hepatic flexure; *T* = transverse colon; *S* = splenic flexure; *D* = descending colon; *G* = sigmoid colon; *R* = rectum; *U* = haustra. Reproduced with permission from Moore, KL: *Clinically Oriented Anatomy*, 3rd edition, Baltimore, Williams & Wilkins, 1992.

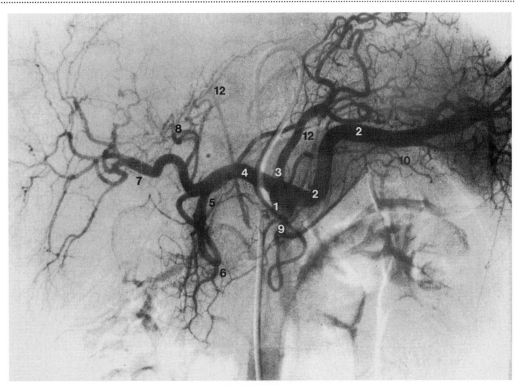

Figure 3-14.   An arteriogram showing the various branches of the celiac trunk. *1* = tip of catheter in the celiac trunk; *2* = splenic artery; *3* = left gastric artery; *4* = hepatic artery; *5* = gastroduodenal artery; *6* = superior pancreaticoduodenal artery; *7* = right hepatic artery; *8* = left hepatic artery; *9* = dorsal pancreatic artery; *10* = left gastroepiploic artery; *11* = right gastroepiploic artery; *12* = phrenic artery; *13* = transverse pancreatic artery; *14* = pancreatic magna artery. Reproduced with permission from Weir and Abrahams: *Imaging Atlas of Human Anatomy*, London, UK, Mosby International, 1992.

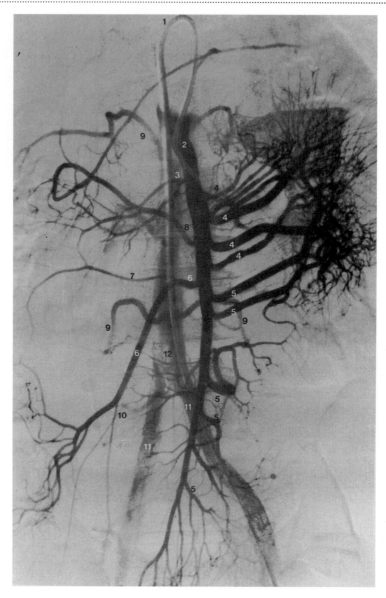

Figure 3-15.   An arteriogram showing the various branches of the superior mesenteric artery. *1* = catheter in superior mesenteric artery; *2* = superior mesenteric artery; *3* = inferior pancreaticoduodenal artery; *4* = jejunal branches of superior mesenteric artery; *5* = ileal branches of superior mesenteric artery; *6* = ileocolic artery; *7* = right colic artery; *8* = middle colic artery; *9* = lumbar arteries; *10* = appendicular artery; *11* = iliac artery; *12* = aorta. Reproduced with permission from Weir and Abrahams: *Imaging Atlas of Human Anatomy*, London, UK, Mosby International, 1992.

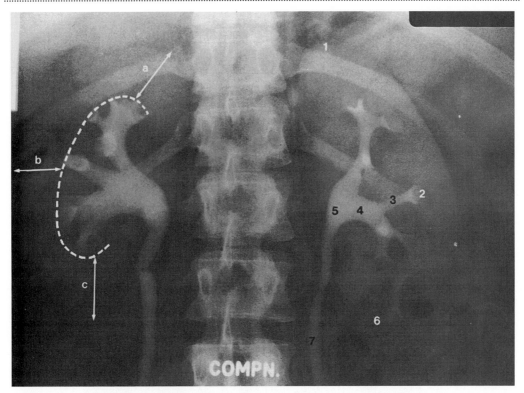

Figure 3-16. A pyelogram of the kidney. The ureter meets the renal pelvis at the ureteropelvic junction, which is a common site of urinary flow obstruction. *a* = upper pole cortex of right kidney; *b* = middle pole cortex of right kidney; *c* = lower pole cortex of right kidney; *1* = upper pole of left kidney; *2* = renal papilla; *3* = minor calyx; *4* = major calyx; *5* = renal pelvis; *6* = lower pole of left kidney; *7* = left ureter. Reproduced with permission from Weir and Abrahams: *Imaging Atlas of Human Anatomy*, London, UK, Mosby International, 1992.

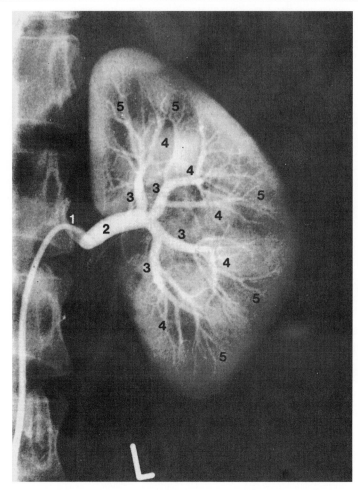

Figure 3-17.   An arteriogram of the kidney. The right renal artery is longer than the left renal artery and lies posterior to the IVC. *1* = tip of catheter in left renal artery; *2* = left renal artery; *3* = lobar arteries; *4* = interlobar arteries *5* = arcuate arteries. Reproduced with permission from Weir and Abrahams: *Imaging Atlas of Human Anatomy*, London, UK, Mosby International, 1992.

4

Pelvis

I. BONY PELVIS (Figure 4-1)

A. **Two coxal (hip) bones, sacrum,** and **coccyx** form the bony pelvis.

    1. Each coxal bone consists of three parts:

        a. Ischium

        b. Ilium

        c. Pubis

    2. These three parts join at the **acetabulum** of the hip joint.

B. Ligaments

    1. The **sacrotuberous ligament** runs from the sacrum to the ischial tuberosity.

    2. The **sacrospinous ligament** runs from the sacrum to the ischial spine.

    3. These ligaments form the borders of the **greater sciatic foramen** and the **lesser sciatic foramen,** through which important structures pass (Table 4- 1).

C. The **pelvic inlet** is defined by the **sacral promontory** (S1 vertebral body) and the **linea terminalis.**

    1. The linea terminalis includes the **pubic crest, iliopectineal line,** and the **arcuate line** of the ilium.

    2. The pelvic inlet divides the pelvis into **two parts:**

        a. The **major (false) pelvis** lies above the pelvic inlet between the iliac crests and is actually part of the abdominal cavity.

        b. The **minor (true) pelvis** lies below the pelvic inlet and extends to the pelvic outlet.

    3. **Measurements** of the pelvic inlet (Figure 4-2)

        a. **Transverse** diameter is the widest distance across the pelvic inlet.

        b. **Oblique** diameter is the distance from the sacroiliac joint to the contralateral iliopectineal eminence.

        c. **True conjugate** diameter is the distance from the sacral promontory to the superior margin of the pubic symphysis. This diameter is measured radiographically on a lateral projection.

        d. **Diagonal conjugate** diameter is the distance from the sacral promontory to the inferior margin of the pubic symphysis. This diameter is measured during an obstetrical examination.

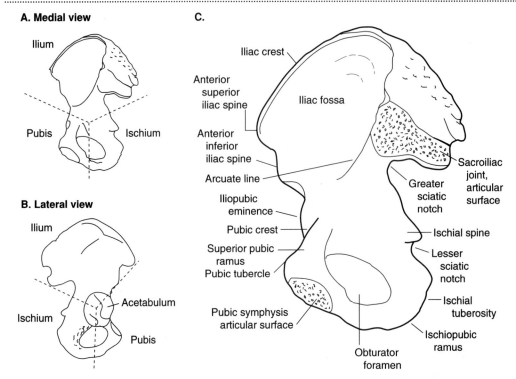

Figure 4-1. (A) Medial view of the coxal (hip) bone, consisting of the ischium, ilium, and pubis. (B) Lateral view of the coxal bone. (C) A schematic diagram of the coxal bone, showing its various anatomical components. Note that in the natural position of the coxal bone, the anterior superior iliac spine and the pubic tubercle lie in the same vertical plane. Adapted with permission from Mathers, LH; Chase, RA; Dolph, J; et al.: *Clinical Anatomy Principles*, St. Louis, Mosby–Yearbook, 1995.

   D. The **pelvic outlet** is defined by the **coccyx, ischial tuberosities, inferior pubic ramus, and pubic symphysis.**

   1. The pelvic outlet is closed by the **pelvic diaphragm** and **urogenital diaphragm.**

   2. Measurements of the pelvic outlet (Figure 4-3)

   a. **Transverse** diameter is the distance between the ischial tuberosities.

   b. **Interspinous** diameter is the distance between the ischial spines. The ischial spines may present a barrier to the fetus during childbirth if the interspinous distance is less than 9.5 cm.

   E. The **birth canal** consists of the **pelvic inlet, minor pelvis, cervix, vagina,** and **pelvic outlet.**

II.   PELVIC DIAPHRAGM. The following muscles comprise the pelvic diaphragm:

   A. **Coccygeus** muscle

   B. **Levator ani** muscles

   1. **Iliococcygeus** muscle

   2. **Pubococcygeus** muscle

Table 4-1.

Structures Traversing the Greater and Lesser Sciatic Foramens

Structures Traversing Greater Sciatic Foramen	Structures Traversing Lesser Sciatic Foramen
Piriformis muscle	
Sciatic nerve	
Superior gluteal vein, artery, nerve	
Inferior gluteal vein, artery, nerve	
Posterior femoral cutaneous nerve	
Nerve to quadratus femoris muscle	
Internal pudendal vein, artery	Internal pudendal vein, artery
Pudendal nerve	Pudendal nerve
Nerve to obturator internus muscle	Nerve and tendon of obturator internus muscle

3. The **puborectalis** muscle forms a U-shaped sling around the anorectal junction at the 90-degree anorectal flexure. This muscle is an important factor in maintaining fecal continence.

III. UROGENITAL DIAPHRAGM. The following muscles comprise the urogenital diaphragm:

A. Deep transverse perineal muscle

B. Sphincter urethra muscle

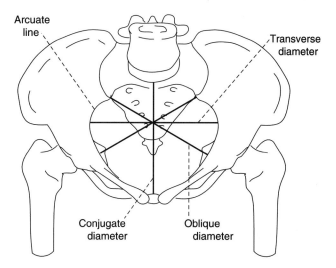

Figure 4-2.   An anterior–superior view of the pelvis, demonstrating the transverse, oblique, and conjugate diameters of the pelvic inlet. These diameters are measured at the level of the linea terminalis (or arcuate line). Adapted with permission from Mathers, LH; Chase, RA; Dolph, J; et al.: *Clinical Anatomy Principles*, St. Louis, Mosby–Yearbook, 1995.

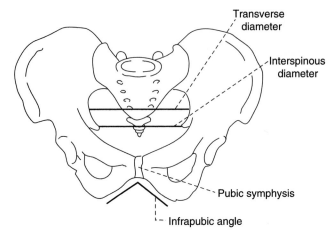

Figure 4-3. An anterior view of a female pelvis, demonstrating the transverse and interspinous diameters of the pelvic outlet. The infrapubic angle in the female pelvis is larger than that in the male. Adapted with permission from Mathers, LH; Chase, RA; Dolph, J; et al.: *Clinical Anatomy Principles*, St. Louis, Mosby–Yearbook, 1995.

IV. URETER (Figure 4-4)

    A. The ureter courses along the lateral wall of the pelvis **retroperitoneally** and **anterior** to the internal iliac arteries and psoas major muscle.

    B. In the **male,** the ureter passes **posterior** to the ductus deferens.

    C. In the **female,** the ureter passes **posterior** to the uterine artery which lies in the transverse cervical ligament (cardinal ligament of Mackenrodt). The ureter may be damaged if it is inadvertently clamped or ligated along with the uterine artery during a hysterectomy.

    D. **Kidney stones** obstruct the ureters most commonly where the ureter:

        1. Crosses the pelvic inlet

        2. Runs obliquely through the wall of the urinary bladder (intramural portion)

V. URINARY BLADDER (Figures 4-5, 4-6, 4-7; see Figure 4-4)

    A. In the **adult,** the empty bladder lies within the **pelvis minor.**

        1. As the bladder fills, it rises out of the pelvis minor above the pelvic inlet and may extend as high as the umbilicus.

        2. In **acute retention** of urine, a needle may be passed through the anterior abdominal wall **without** entering the peritoneal cavity in order to drain off the urine. The needle passes through the following structures:

            a. Skin

            b. Superficial fascia (Camper's and Scarpa's)

            c. Linea alba

            d. Transversalis fascia

            e. Extraperitoneal fat

            f. Wall of the urinary bladder

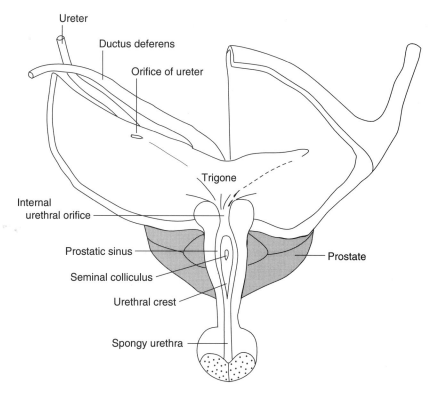

Figure 4-4.   The interior of the male urinary bladder and prostatic urethra. The anterior portions of the bladder, prostate, and urethra are cut away. The posterior wall of the bladder contains the interureteric fold, which unites the ureters along the superior border of the trigone. Note the anatomical relationship of the ureter to the ductus deferens. Adapted with permission from Moore, KL: *Clinically Oriented Anatomy*, 3rd edition, Baltimore, Williams & Wilkins, 1992.

B. In the **infant,** the empty bladder lies within the **abdominal cavity.**

C. The **anterior surface** of the bladder is related to the **pubic symphysis** and the **retropubic space.** The retropubic space can be used as a surgical approach to the prostate gland.

D. The **posterior surface** of the bladder is related to the:

  1. Rectovesical pouch, seminal vesicles, and ampulla of ductus deferens in **males**

  2. Vesicouterine pouch and anterior wall of the vagina in **females**

E. The **base of the urinary bladder** is related to the prostate gland in the male.

F. The **urinary bladder** is related to various **ligaments:**

  1. Median umbilical ligament (there is one) is a remnant of the urachus (or allantois) in the embryo.

  2. Medial umbilical ligaments (there are two) are remnants of the right and left umbilical arteries in the embryo.

  3. Lateral umbilical ligaments (there are two) are defined by the inferior epigastric artery and vein.

G. If the **superior wall** of the urinary bladder is **ruptured** because of an injury to the abdominal wall, the peritoneum reflected over its surface frequently is torn, causing urine to leak into the peritoneal cavity (intraperitoneal).

H. If the **anterior wall** of the urinary bladder is ruptured because of a fractured pelvis, urine will leak into the retropubic space (extraperitoneal).

VI. SIGMOID COLON

A. The sigmoid colon is a segment of large intestine between the descending colon and the rectum whose primary function is **storage of feces.**

B. The sigmoid colon begins at the pelvic inlet and ends at vertebral level S3.

C. The sigmoid colon is suspended by the sigmoid mesocolon (i.e., intraperitoneal). The **left ureter** and the **bifurcation of the left common iliac artery** lie at the apex of the sigmoid mesocolon.

D. **Arterial supply** is via four **sigmoidal arteries** and the **rectosigmoid artery,** which are branches of the inferior mesenteric artery.

E. **Venous return** is via the inferior mesenteric vein into the hepatic portal system.

F. During a **sigmoidoscopy,** the large intestine may be punctured if the angle at the rectosigmoid junction is not negotiated properly. At the rectosigmoid junction, the sigmoid colon bends in an anterior direction and to the left.

G. The sigmoid colon is often used in a **colostomy** due to the mobility rendered by its mesentery.

VII. RECTUM (see Figures 4-5, 4-6)

A. The rectum is a segment of large intestine between the sigmoid colon and the anal canal.

B. The rectum joins the anal canal at a 90-degree angle, forming the **anorectal flexure.** The puborectalis muscle forms a U-shaped sling at the anorectal junction and maintains the 90-degree angle.

C. The **ampulla** of the rectum lies just above the pelvic diaphragm and generates the urge to defecate when feces move into it.

D. **Three transverse rectal folds (Houston's valves)** formed by the mucosa, submucosa, and inner circular layer of smooth muscle **permanently** extend into the lumen of the rectum. The transverse rectal folds need to be negotiated during a proctoscopy or sigmoidoscopy.

E. The **arterial supply** and **venous drainage** are indicated in Table 4-2.

F. **Prolapse** of the rectum through the anus generally is a result of trauma due to childbirth or loss of muscle tone in the aged.

G. **Carcinoma of the rectum** may metastasize to the liver because the superior rectal vein drains into the hepatic portal system. In addition, carcinoma of the rectum may metastasize to nearby surrounding structures.

1. Posterior metastasis—sacral nerve plexus causing sciatica

2. Lateral metastasis—ureter

3. Anterior metastasis

a. Male—prostate, seminal vesicle, and urinary bladder

b. Female—uterus and vagina

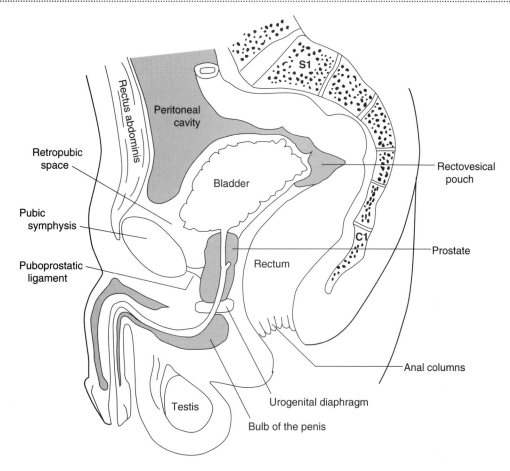

Figure 4-5. A sagittal section through the male pelvis, demonstrating various anatomical relationships of pelvic viscera. Adapted with permission from Moore, KL: *Clinically Oriented Anatomy*, 3rd edition, Baltimore, Williams & Wilkins, 1992.

VIII.   PROSTATE (see Figures 4-4, 4-5, 4-7)

   A. The prostate is located between the base of the urinary bladder and the pelvic diaphragm.

   B. The **anterior surface** of the prostate is related to the retropubic space.

   C. The **posterior surface** of the prostate is related to the seminal vesicles and rectum. The prostate gland can be easily palpated by a digital examination via the rectum.

   D. The prostate consists of **five lobes.**

   1. Right and left **lateral** lobes

   2. Right and left **posterior** lobes, which are predisposed to **malignant transformation** (the most common malignancy in adult males)

   3. **Median** lobe

   a. This lobe is predisposed to **benign prostatic hypertrophy.**

   b. The **hypertrophied median lobe** encroaches on the sphincter vesicae, located at the neck of the bladder, and allows leakage of urine into the prostatic urethra. This leakage causes a reflex desire to micturate.

Table 4-2.

Arterial Supply and Venous Drainage of the Rectum

Rectum	Arterial Supply	Venous Drainage
Upper portion	Superior rectal artery* (branch of the inferior mesenteric artery)	Superior recal vein → inferior mesenteric vein → hepatic portal system
Middle portion	Middle rectal artery (branch of internal iliac artery)	Middle rectal → internal iliac vein → IVC
Lower portion	Inferior rectal artery (branch of internal pudendal artery)	Inferior rectal vein → internal pudendal vein → internal iliac vein → IVC

* The superior rectal artery is the chief blood supply to the rectum.
IVC = inferior vena cava

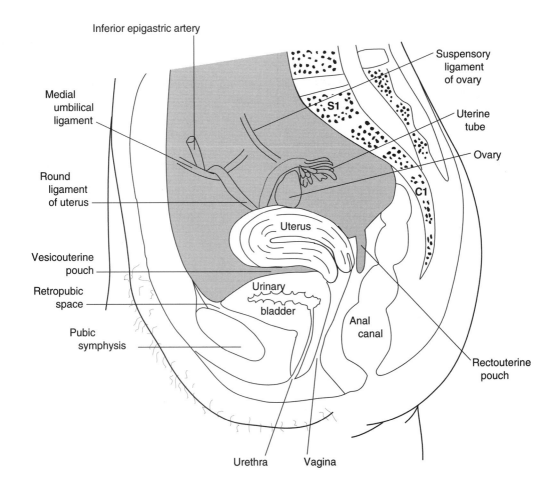

Figure 4-6. A sagittal section through the female pelvis, demonstrating various anatomical relationships of the pelvic viscera. Adapted with permission from Moore, KL: *Clinically Oriented Anatomy*, 3rd edition, Baltimore, Williams & Wilkins, 1992.

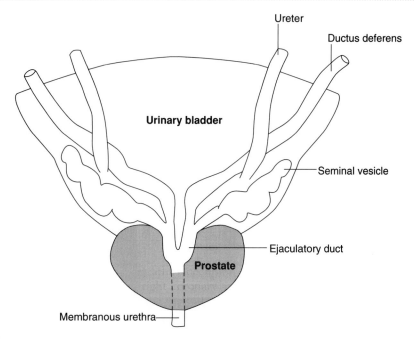

Figure 4-7. A posterior view of the urinary bladder, demonstrating its anatomical relationship to the seminal vesicle, ductus deferens, and prostate gland. Adapted with permission from Moore, KL: *Clinically Oriented Anatomy*, 3rd edition, Baltimore, Williams & Wilkins, 1992.

  E. Prostatic fluid contains **citric acid, acid phosphatase, prostaglandins, fibrinogen,** and **prostate-specific antigen (PSA).** Serum levels of acid phosphatase and PSA are used as diagnostic tools for **prostatic cancer.**

  F. The **prostatic venous plexus** drains into the:

   1. Internal iliac veins, then to the **inferior vena cava (IVC),** which may explain the metastasis of prostatic cancer to the heart and lungs

   2. Vertebral venous plexus, which may explain the metastasis of prostatic cancer to the vertebral column and brain

IX. UTERUS (Figure 4-8)

  A. The uterus is divided into **four regions.**

   1. The **fundus** is located superior to the cornua and contributes largely to the upper segment of the uterus during pregnancy. At term, the fundus may extend as high as the xiphoid process (T9).

   2. The **cornu** defines the entry of the uterine tube.

   3. The **body** is located between the cornu and the cervix. The **isthmus** is part of the body and is the dividing line between the body of the uterus and the cervix. The isthmus corresponds to the internal os and is the preferred site for cesarean section.

   4. The **cervix** projects into the vagina and is divided into **three parts:**

    a. Internal os

    b. Cervical canal

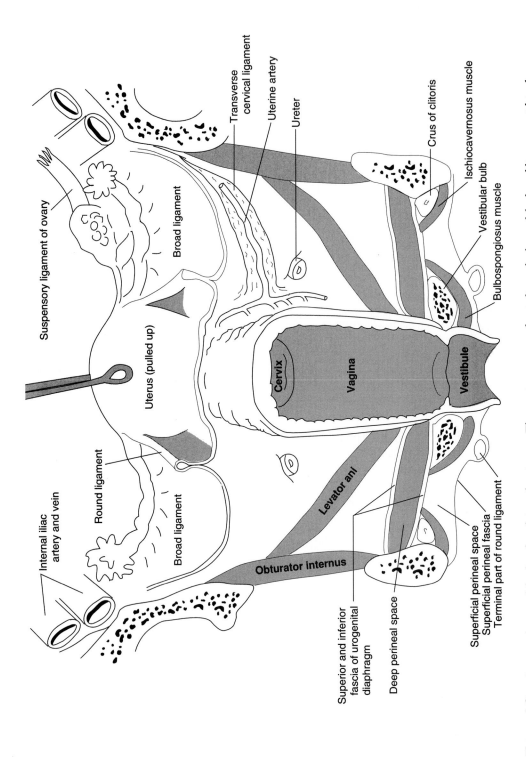

Figure 4-8. A coronal view of the female pelvis and perineum. The uterine artery and vein lie at the base of the broad ligament within the transverse cervical ligament. Note the close relationship of the ureter to the uterine artery and vein. Note the uterus/cervix and their relationship to the vagina. Note the various components of the broad ligament. Adapted with permission from Netter, FH: *Atlas of Human Anatomy*, Summit, NJ, Ciba-Geigy Corp., 1989.

    c. External os

        (1) The external os in a **nulliparous** woman is **round.**

        (2) The external os in a **parous** woman is **transverse.**

B. The uterus is supported by the:

    1. Pelvic diaphragm, urogenital diaphragm, and **urinary bladder**

    2. **Round ligament,** which is a remnant of the gubernaculum in the embryo

    3. **Transverse cervical ligament (cardinal ligament of Mackenrodt),** which contains the uterine artery and vein

    4. Uterosacral ligament

    5. Pubocervical ligament

    6. The uterus is normally in an **anteflexed** and **anteverted** position.

        a. **Flexion** refers to the angle between the cervix and body of the uterus.

        b. **Version** refers to the angle between the vagina and cervix.

X. VAGINA

A. The vagina extends from the cervix of the uterus to the vestibule of the vagina.

B. The vagina is the **longest part** of the birth canal, and its distention during childbirth is limited by the **ischial spines** and **sacrospinous ligaments.**

C. The vagina forms a recess around the cervix called the **fornix.** The fornix is divided into the **anterior, posterior,** and **lateral fornix.**

D. The **posterior fornix** is related to the **rectouterine pouch** of the peritoneal cavity and can be used in the following ways:

    1. As a route to insert a culdoscope into the peritoneal cavity to examine the uterine tubes and ovaries

    2. As a route to insert a needle into the peritoneal cavity to collect oocytes for in vitro fertilization

XI. BROAD LIGAMENT (see Figure 4-8)

A. The broad ligament **consists** of the:

    1. **Mesosalpinx,** which supports the uterine tube

    2. **Mesovarium,** which supports the ovary

    3. **Mesometrium,** which supports the uterus

    4. **Suspensory ligament of the ovary,** which is a lateral extension of the broad ligament and contains the ovarian artery, vein, and nerves

B. The broad ligament **contains** the following structures:

    1. **Ovarian artery, vein,** and **nerves**

    2. Uterine tubes

    3. **Ovarian ligament**—remnant of the gubernaculum in the embryo

    4. **Round ligament** of the uterus—remnant of the gubernaculum in the embryo

    5. **Epoophoron**—remnant of the mesonephric tubules in the embryo

6. **Paroöphoron**—remnant of the mesonephric tubules in the embryo

7. **Gartner's duct**—remnant of the mesonephric duct in the embryo

8. **Ureter**—lies at the base of the broad ligament

9. **Uterine artery, vein,** and **nerves**—lie at the base of the broad ligament

XII. CROSS-SECTIONAL ANATOMY

A. Male pelvis at the level of the prostate gland (Figure 4-9)

B. Female pelvis at the level of the cervix (Figure 4-10)

XIII. RADIOLOGY

A. AP radiograph of the bony pelvis (Figure 4-11)

B. Pelvic arteriogram (Figure 4-12)

C. Seminal vesiculogram (Figure 4-13)

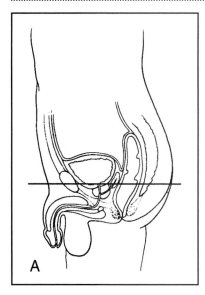

KEY
AC acetabulum
BL urinary bladder
CCX coccyx
CR hyaline articular cartilage
EA external iliac artery
EV external iliac vein
F head of femur
FA femoral artery
FN femoral nerve
FT fat
FV femoral vein
GMA gluteus maximus
GME gluteus medius
GMI gluteus minimus
GT greater trochanter of femur
ICG iliococcygeus
IGA inferior gluteal artery
ILC iliacus
IPS iliopsoas
IPV internal pudendal vessels
IS ischium
NC natal cleft
OE obturator externus

OI obturator internus
OM obturator membrane
OVN obturator vessels and
    nerves
P pubic bone
PEC pectineus
PF perirectal fat
PM psoas major
PR prostate gland
PRV prostatic venous plexus
PS pubic symphysis
PU prostatic urethra
PVN pudendal vessels and
    nerves
R rectum
RAB rectus abdominis
RF rectus femoris
SN sciatic nerve
SPD spermatic card
SR sartorius
SV seminal vesicle
TFL tensor fasciae latae
VD vas deferens

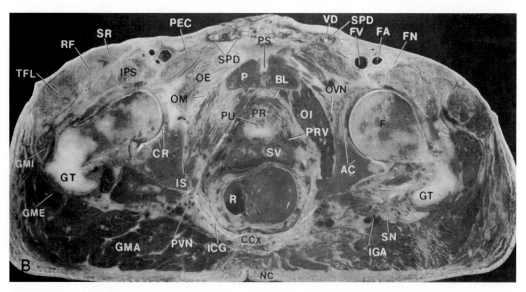

Figure 4-9. A cross section of the male pelvis at the level of the prostate gland. (A) A schematic diagram to show where the cross section was taken. (B) A cross section through a cadaver. The ductus deferens is shown within the spermatic cord just anterior to the pectineus muscle. Note the relationship of the femoral vein, femoral artery, and femoral nerve to the iliopsoas and pectineus muscles. The sciatic nerve lies posteromedial to the greater trochanter. Note the location of the pudendal artery, vein, and nerve. Reproduced with permission from Barrett, CP ; Poliakoff, SJ; Holder, LE; et al.: *Primer of Sectional Anatomy with MRI and CT Correlation*, 2nd edition, Baltimore, Williams & Wilkins, 1994.

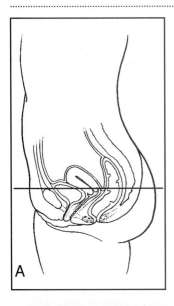

KEY
A anus
AC acetabulum
AF acetabular fat
BL urinary bladder
CCX coccyx
CX cervix of uterus
F head of femur
FA femoral artery
FN femoral nerve
FV femoral vein
GMA gluteus maximus
GT greater trochanter
IPS iliopsoas
IR ischiorectal fossa

IS ischial spine
NC natal cleft
NF neck of femur
OI obturator internus
OVN obturator nerves and vessels
PD pouch of Douglas
PEC pectineus
PVN pudendal vessels and nerves
R rectum
RF rectus femoris
RL round ligament of uterus
SN sciatic nerve
SR sartorius
TFL tensor fasciae latae
VG vagina

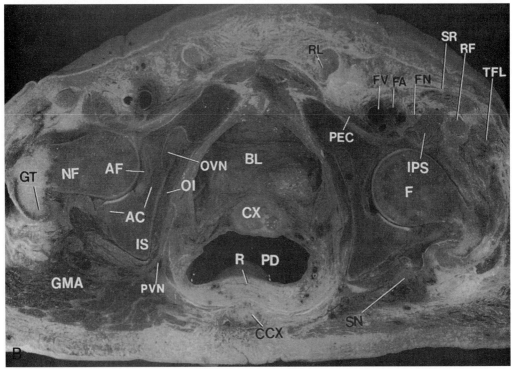

Figure 4-10. A cross section of the female pelvis at the level of the cervix. (A) A schematic diagram to show where the cross section was taken. (B) A cross section through a cadaver. The pudendal artery, vein, and nerve lie posterior to the ischial spine. Note the location of the sciatic nerve, femoral vein, femoral artery, and femoral nerve. The round ligament of the uterus lies in an anterior location. Note the pouch of Douglas (rectouterine pouch) between the rectum and uterus. Reproduced with permission from Barrett, CP ; Poliakoff, SJ; Holder, LE; et al.: *Primer of Sectional Anatomy with MRI and CT Correlation*, 2nd edition, Baltimore, Williams & Wilkins, 1994.

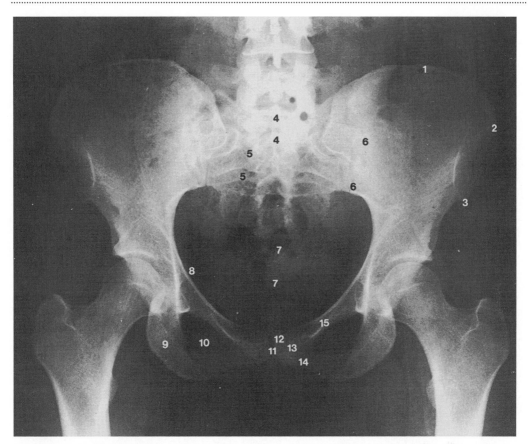

Figure 4-11.   AP radiograph of the bony pelvis. (*1*) Iliac crest; (*2*) anterior superior iliac spine; (*3*) anterior inferior iliac spine; (*4*) sacral crest; (*5*) anterior sacral foramen; (*6*) sacroiliac joint; (*7*) coccyx; (*8*) ischial spine; (*9*) ischial ramus; (*10*) obturator foramen; (*11*) pubic symphysis; (*12*) pubic tubercle; (*13*) pubic body; (*14*) inferior ramus of the pubis; (*15*) superior ramus of the pubis. Reproduced with permission from Weir and Abrahams: *Imaging Atlas of Human Anatomy*, London, UK, Mosby International, 1992.

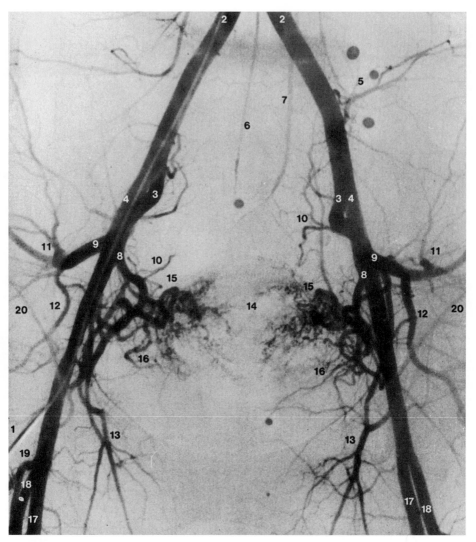

Figure 4-12.   A pelvic arteriogram. (*1*) Catheter in right femoral artery; (*2*) common iliac artery; (*3*) internal iliac artery; (*4*) external iliac artery; (*5*) iliolumbar artery; (*6*) median sacral artery; (*7*) inferior mesenteric artery; (*8*) anterior trunk of internal iliac artery; (*9*) posterior trunk of internal iliac artery; (*10*) lateral sacral artery; (*11*) superior gluteal artery; (*12*) inferior gluteal artery; (*13*) obturator artery; (*14*) position of the uterus; (*15*) uterine artery; (*16*) superior vesical artery; (*17*) femoral artery; (*18*) profunda femoris artery; (*19*) lateral circumflex femoral artery; (*20*) deep circumflex iliac artery. Reproduced with permission from Weir and Abrahams: *Imaging Atlas of Human Anatomy*, London, UK, Mosby International, 1992.

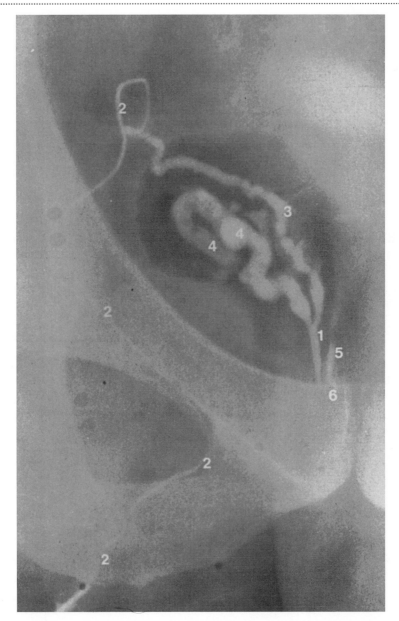

Figure 4-13.   A seminal vesiculogram. (*1*) Right ejaculatory duct; (*2*) ductus deferens; (*3*) ampulla of ductus deferens; (*4*) seminal vesicle; (*5*) left ejaculatory duct; (*6*) position of the seminal colliculus. Reproduced with permission from Weir and Abrahams: *Imaging Atlas of Human Anatomy*, London, UK, Mosby International, 1992.

5

Perineum

I. PERINEUM (Figure 5-1). The perineum is a diamond-shaped region inferior to the pelvic diaphragm. It can be divided by a line passing through the ischial tuberosities into the **anal** and **urogenital (UG) triangles.**

A. The **anal triangle** contains the **anal canal,** which is divided into the **upper** and **lower anal canal** by the **pectinate line.** Major distinctions between the upper and lower anal canal are shown in Table 5-1.

B. The **urogenital triangle** contains the outlets of the urinary and genital systems.

II. SUPERFICIAL PERINEAL SPACE (Figure 5-2)

A. Boundaries of this space include the **superficial perineal fascia** (which is continuous with Colles' fascia) and the **inferior fascia** of the **UG diaphragm** (also called the **perineal membrane**).

B. Structures of the superficial perineal space in both male and female are shown in Table 5-2.

C. Rupture of the male urethra (straddle injuries) allows urine to escape into the superficial perineal space and pass into the connective tissue around the **scrotum, penis,** and **anterior abdominal wall.** Urine will **not** pass into the **thigh region** or **anal triangle.**

III. DEEP PERINEAL SPACE (see Figure 5-2)

A. Boundaries of this space include the **inferior and superior fasciae** of the **UG diaphragm.**

B. Structures of the deep perineal space in both male and female are shown in Table 5-3.

IV. MALE GENITALIA (Figure 5-3)

A. The **penis** consists of **three columns** of erectile tissue that are bound together by the **tunica albuginea.**

   1. Corpus spongiosum (one column) transmits the urethra. Proximally, the corpus spongiosum enlarges to form the **bulb of the penis** and distally enlarges to form the **glans of the penis.**

   2. Corpus cavernosa (two columns)

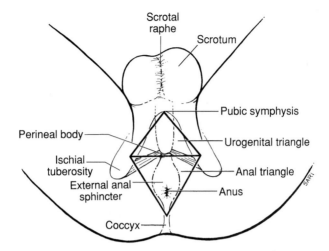

Figure 5-1.   The diamond-shaped perineum, extending from the pubic symphysis to the coccyx. Note that a transverse line joining the anterior ends of the ischial tuberosities divides the perineum into two unequal triangular areas, the urogenital triangle anteriorly and the anal triangle posteriorly. The midpoint of the transverse line indicates the site of the perineal body (central perineal tendon). Reproduced with permission from Moore, KL: *Clinically Oriented Anatomy*, 3rd edition, Baltimore, Williams & Wilkins, 1992.

    B. The **scrotum** consists of **three parts:**

        1. Skin

        2. Colles' fascia

        3. Dartos muscle

    C. Testes

Table 5-1.

Distinctions Between the Upper and Lower Anal Canal

Feature	Upper Anal Canal	Lower Anal Canal
Arterial supply	Superior rectal artery (branch of inferior mesenteric artery)	Inferior rectal artery (branch of internal pudendal artery)
Venous drainage	Superior rectal vein → inferior mesenteric vein → hepatic portal system	Inferior rectal vein → internal pudendal vein → internal iliac vein → IVC
Lymphatic drainage	Deep nodes	Superficial inguinal nodes
Innervation	Motor: internal anal sphincter (smooth muscle); autonomic Sensory: stretch sensation	Motor: external anal sphincter (striated muscle); pudendal nerve Sensory: pain, temperature, touch
Embryological derivation	Endoderm (hindgut)	Ectoderm (proctodeum)
Epithelium	Simple columnar	Stratified squamous
Hemorrhoids	Internal hemorrhoids (varicosities of the superior rectal vein)	External hemorrhoids (varicosities of the inferior rectal vein

IVC = inferior vena cava

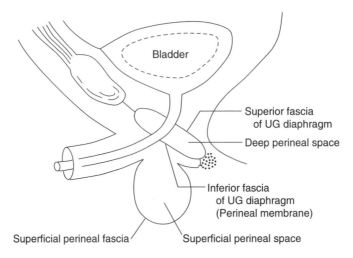

Figure 5-2. A median section of the male pelvis showing the superficial perineal space and deep perineal space, along with their fascial boundaries. Adapted with permission from Moore, KL: *Clinically Oriented Anatomy*, 3rd edition, Baltimore, Williams & Wilkins, 1992.

V. FEMALE GENITALIA

A. The **clitoris** consists of **three parts** similar to those in the penis except that the clitoris has no corpus spongiosum and does not transmit the urethra.

 1. Crura of the clitoris

 2. Corpus cavernosa

 3. Glans of the clitoris

B. Labia majora

C. Labia minora

Table 5-2.

Structures of the Superficial Perineal Space

Structure	Male	Female
Erectile tissue	Bulb of the penis Crura of the penis	Vestibular bulbs Crura of the clitoris
Organs	Urethra	Urethra Vagina
Muscle	Bulbospongiosus Ischiocavernosus Superficial transverse perineal	Bulbospongiosus Ischiocavernosus Superficial transverse perineal
Connective tissue	Perineal body	Perineal body
Nerve (Branch of pudendal nerve)	Perineal nerve	Perineal nerve
Glands	None	Greater vestibular (Bartholin's)

Table 5-3.

Structures of the Deep Perineal Space

Structure	Male	Female
Erectile tissue	None	None
Organs	Urethra	Urethra Vagina
Muscle	Deep transverse perineal Sphincter urethrae	Deep transverse perineal Sphincter urethrae
Connective tissue	None	None

Nerve
  (Branch of pudendal
   nerve)
	Perineal nerve
	Dorsal nerve of the penis
		Perineal nerve
		Dorsal nerve of the clitoris

Glands	Bulbourethral (Cowper's)	None

D. Vestibule is the space between the labia minora into which the **vagina, urethra, paraurethral glands,** and **greater vestibular glands** open.

E. Episiotomy (Figure 5-4). When it appears inevitable that the perineum or perineal body will tear during childbirth, an incision is made to enlarge the vaginal orifice.

    1. Median episiotomy. This incision starts at the **frenulum of the labia minora** and extends through the following structures.

        a. Skin

        b. Vaginal wall

        c. Perineal body

        d. Superficial transverse perineal muscle

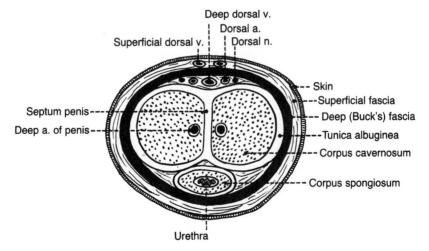

Figure 5-3.   Cross section of the penis. Reproduced with permission from Chung, KW: *BRS Gross Anatomy*, 2nd edition, Baltimore, Williams & Wilkins, 1991.

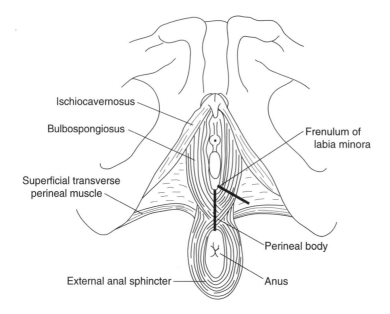

Figure 5-4. The female perineum in the lithotomy position showing the incision lines for a median and mediolateral episiotomy. Adapted with permission from Moore, KL: *Clinically Oriented Anatomy*, 3rd edition, Baltimore, Williams & Wilkins, 1992.

2. Mediolateral episiotomy. This incision starts at the **frenulum of the labia minora** and extends through the following structures.

   a. Skin

   b. Vaginal wall

   c. Bulbospongiosus muscle

F. Pudendal nerve block (Figure 5-5)

   1. In order to relieve pain associated with childbirth, the **pudendal nerve** may be blocked via the perineal route by using the **ischial tuberosity** as the chief bony landmark.

   2. When complete perineal anesthesia is required, the **genitofemoral nerve, ilioinguinal nerve,** and **perineal branch of the posterior cutaneous nerve of the thigh** must also be blocked by making injections along the lateral margin of the **labia majora.**

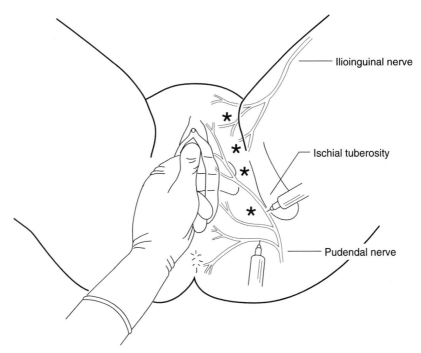

Ilioinguinal nerve

Ischial tuberosity

Pudendal nerve

Figure 5-5.   The female perineum in the lithotomy position showing the site of injection for a pudendal nerve block using the ischial tuberosity as a landmark. For complete anesthesia of the area, additional injections along the lateral border of the labia majora are performed (*). Adapted with permission from Moore, KL: *Clinically Oriented Anatomy*, 3rd edition, Baltimore, Williams & Wilkins, 1992.

6

Upper Limb

I. ARTERIAL SUPPLY (Figure 6-1)

A. The **subclavian artery** extends from the arch of the aorta to the lateral border of the first rib. The subclavian artery branches off into the following arteries:

    1. The **internal thoracic artery** is continuous with the **superior epigastric artery,** which anastomoses with the **inferior epigastric artery** (a branch of the external iliac artery). This may provide a route of collateral circulation if the abdominal aorta is blocked.

    2. Vertebral artery

    3. The **thyrocervical trunk** has three branches, including:

        a. **Suprascapular artery,** which participates in collateral circulation around the shoulder

        b. **Transverse cervical artery,** which participates in collateral circulation around the shoulder

        c. Inferior thyroid artery

    4. Costocervical trunk

B. The **axillary artery** is a continuation of the subclavian artery and extends from the lateral border of the first rib to the inferior border of the teres major muscle. The tendon of the pectoralis minor muscle crosses the axillary artery anteriorly. The axillary artery branches into the following arteries:

    1. Supreme thoracic artery

    2. Thoracicoacromial artery

    3. Lateral thoracic artery

    4. Posterior humeral circumflex artery

    5. Anterior humeral circumflex artery

    6. Subscapular artery

C. The **brachial artery** is a continuation of the axillary artery and extends from the lateral border of the teres major muscle to the cubital fossa. The brachial artery branches off into the following arteries:

    1. **Deep brachial artery.** A fracture of the humerus at midshaft may damage the **deep brachial artery** and **radial nerve** as they travel together on the posterior aspect of the humerus in the radial groove.

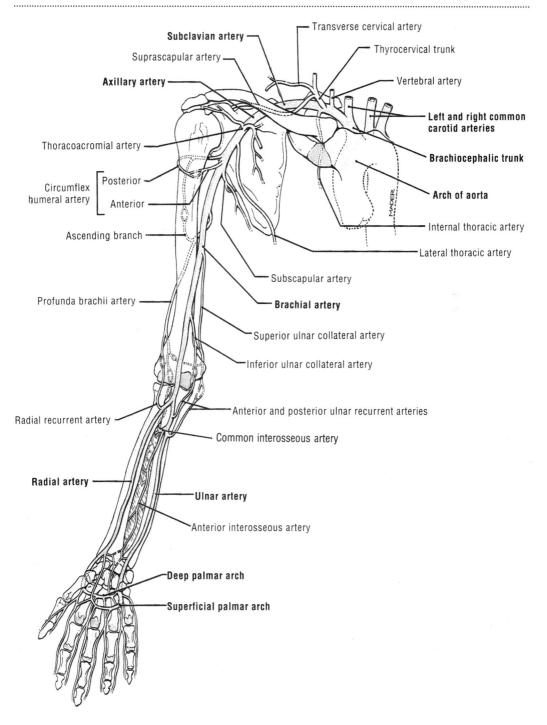

Figure 6-1.   Arterial supply of the upper limb. Adapted with permission from Moore, KL: *Clinically Oriented Anatomy*, 3rd edition, Baltimore, Williams & Wilkins, 1992.

    2. Superior ulnar collateral artery

    3. Inferior ulnar collateral artery

D. Radial artery

E. Ulnar artery

F. Superficial palmar arch

G. The **deep palmar arch** lies posterior to the tendons of the flexor digitorum superficialis and flexor digitorum profundus muscles. A **deep laceration** at the metacarpal–carpal joint that cuts the deep palmar arch compromises flexion of the fingers.

H. Collateral circulation

    1. Around the shoulder (Figure 6-2)

        a. Transverse cervical artery

        b. Suprascapular artery

        c. Subscapular artery

    2. Around the elbow

        a. Superior ulnar collateral artery anastomoses with the **posterior ulnar recurrent artery.**

        b. Inferior ulnar collateral artery anastomoses with the **anterior ulnar recurrent artery.**

        c. Middle collateral artery anastomoses with the **interosseous recurrent artery.**

        d. Radial collateral artery anastomoses with the **radial recurrent artery.**

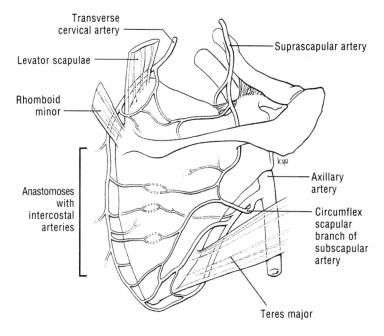

Figure 6-2.   A posterior view showing the arterial anastomoses around the scapula. Adapted with permission from Moore, KL: *Clinically Oriented Anatomy,* 3rd edition, Baltimore, Williams & Wilkins, 1992.

3. In the hand

   a. Superficial palmar arch

   b. Deep palmar arch

I. Placement of ligatures

   1. A surgical ligature may be placed in the following locations:

   a. On the subclavian artery or axillary artery **between the thyrocervical trunk** and **subscapular artery**

   b. On the brachial artery **distal to the inferior ulnar collateral artery**

   2. A surgical ligature may **not** be placed on the axillary artery just **distal to the sub-scapular artery.**

J. Compartment syndrome

   1. Both the upper and lower limbs are separated into **compartments** by various **fascial sheets.**

   2. A **traumatic injury** to the limb may cause progressive **hemorrhage** into one of these compartments, which compresses uninjured blood vessels and/or nerves, producing **is-chemia** and **atrophy of the musculature.** Muscle movements are painful and weak-ened when performed against resistance.

II. BRACHIAL PLEXUS (Figure 6-3). The subdivisions of the brachial plexus include:

A. Rami are the **C5–T1 ventral primary rami** of spinal nerves and are located between the **anterior** and **middle scalene muscles.**

B. Trunks (upper, middle, lower) are formed by the joining of rami and are located in the **posterior triangle of the neck.**

C. Divisions (three anterior, three posterior) are formed by **trunks dividing** into anterior and posterior divisions and are located **deep to the clavicle.**

D. Cords (lateral, medial, posterior) are formed by joining of the anterior and posterior divisions, are located in the **axilla deep to the pectoralis minor muscle,** and are named according to their relationship to the **axillary artery.**

E. Branches. The **five major terminal branches** are:

   1. Musculocutaneous nerve

   2. Axillary nerve

   3. Radial nerve

   4. Median nerve

   5. Ulnar nerve

F. Injuries to the brachial plexus (Table 6-1)

III. NERVE LESIONS

A. Axillary nerve injury may be caused by a fracture of the **surgical neck of the humerus** or **dislocation of the shoulder joint.**

   1. Paralysis of the **deltoid muscle** occurs so that **abduction of the arm to the horizontal position** is compromised.

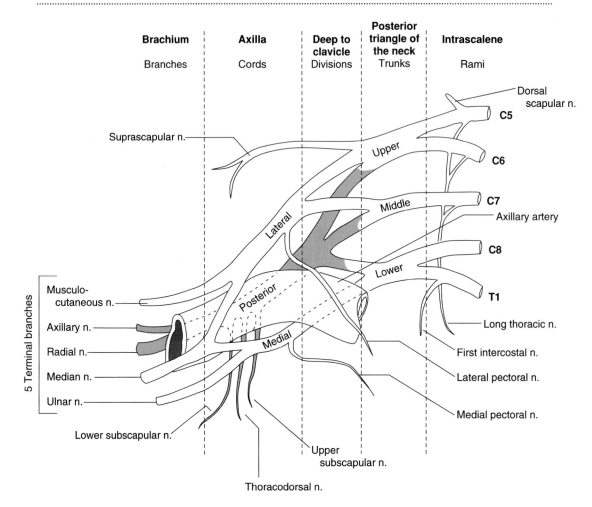

Figure 6-3.   A diagram of the brachial plexus showing the rami, trunks, divisions, cords, and branches along with their respective anatomical position. A portion of the axillary artery and its relationship to the lateral, medial, and posterior cords is shown. Adapted with permission from April, EW: NMS *Clinical Anatomy*, 3rd edition, Baltimore, Williams & Wilkins, 1997.

> **2.** Paralysis of the **teres minor muscle** occurs so that **lateral rotation of the arm** is weakened.
>
> **3.** Sensory loss occurs on the **lateral side of the upper arm.**
>
> **4.** To **test** the deltoid muscle clinically, the patient's arm is abducted to the horizontal and then the patient holds that position against a downward pull.
>
> **B. Long thoracic nerve injury** may be caused by a **stab wound** or **removal of lymph nodes** during a mastectomy.
>
> > **1.** Paralysis of the **serratus anterior muscle** occurs so that **abduction of the arm past the horizontal position** is compromised. In addition, the arm cannot be used to push.
> >
> > **2.** To test the serratus anterior muscle clinically, the patient is asked to face a wall and push against it with both arms. The medial border and inferior angle of the scapula on the injured side become prominent (called **winging of the scapula**).

Table 6-1.

Injuries to the Brachial Plexus

Injury	Cervical Level	Injury Description	Nerves Damaged	Muscles Affected	Clinical Sign
Erb-Duchenne (upper trunk)	C5 and C6	Violent stretch between the head and shoulder	Musculocutaneous	Biceps brachii Brachialis	Arm is medially rotated and pronated
			Suprascapular	Infraspinatus	(Waiter's tip)*
			Axillary	Teres minor	
Klumpke's (lower trunk)	C8 and T1	Sudden pull upward of arm	Median Ulnar	Muscles of the hand and wrist	Loss of function of the hand and wrist

* The infraspinatus muscle, innervated by the suprascapular nerve, is a lateral rotator of the arm. If the infraspinatus muscle is weakened, the medial rotators of the arm dominate. The biceps brachii muscle, innervated by the musculocutaneous nerve, assists in supination. If the biceps brachii muscle is weakened, the pronators of the arm dominate.

C. Radial nerve injury may be caused by a fracture of the **humerus at midshaft.**

   1. Paralysis of the **muscles in the extensor compartment of the forearm** occurs so that **extension of the wrist and digits** is lost and **supination** is compromised. **Extension of the forearm** is preserved because innervation to the triceps muscle is generally intact.

   2. Sensory loss occurs on the **posterior arm, posterior forearm,** and **lateral aspect of the dorsum of the hand.**

   3. Clinically, the hand is flexed at the wrist and lies flaccid in a condition known as **wrist-drop.**

D. Median nerve injury at the elbow or axilla may be caused by a **supracondylar fracture of the humerus.**

   1. Paralysis of the **muscles in the flexor compartment** of the arm occurs so that **flexion of the wrist** is weakened, **flexion of the digits** is lost, and **pronation** is lost.

   2. Paralysis of **abductor pollicis brevis, opponens pollicis,** and **flexor pollicis brevis muscles** occurs so that **thumb movements** are lost.

   3. Sensory loss occurs on the **palmar and dorsal aspects of the index, middle, and half of the ring fingers** and the **palmar aspect of the thumb.**

   4. Clinically, a flattening of the thenar eminence occurs in a condition known as **ape hand.**

E. Median nerve injury at the wrist may be caused by a **slashing of the wrist** or **carpal tunnel syndrome** (see VII B).

   1. There is no paralysis of the muscles in the flexor compartment of the arm.

   2. Paralysis of **abductor pollicis brevis, opponens pollicis,** and **flexor pollicis brevis muscles** occurs so that **thumb movements** are lost.

   3. Sensory loss occurs on the **palmar and dorsal aspects of the index, middle, and half of the ring fingers** and the **palmar aspect of the thumb.**

   4. Clinically, **ape hand** is observed (see III D 4).

F. Ulnar nerve injury at the elbow or axilla may be caused by a fracture of the **medial epicondyle of the humerus.**

1. Paralysis of **flexor carpi ulnaris muscle** occurs so that **deviation of the hand to the radial side upon flexion** at the wrist joint is observed.

2. Paralysis of the **medial part of flexor digitorum profundus muscle** occurs so that **flexion of the ring and little fingers at the distal phalangeal joint** is lost.

3. Paralysis of **lumbricales 3 and 4** occurs so that **flexion of the ring** and **little fingers at the metacarpophalangeal joint** is lost. However, the extensor muscles are intact so that there is hyperextension at these joints.

4. Paralysis of **palmar and dorsal interosseous muscles** occurs so that **abduction and adduction of the fingers** are lost.

5. Paralysis of the **adductor pollicis muscle** occurs so that **adduction of the thumb** is lost.

6. Paralysis of the **abductor digiti minimi, flexor digiti minimi,** and **opponens digiti minimi muscles** occurs so that **abduction, flexion, and opposition of the little finger** are lost.

7. Sensory loss occurs on the **palmar and dorsal aspects of the little finger and half of the ring finger.**

8. Clinically, because of the paralysis of the medial part of the flexor digitorum profundus muscle and lumbricales 3 and 4, the hand has a distinctive shape called **clawhand.**

G. **Ulnar nerve injury at the wrist** may be caused by a **slashing of the wrist.**

1. There is no paralysis of flexor carpi ulnaris muscle.

2. There is no paralysis of medial part of the flexor digitorum profundus muscle.

3. Paralysis of **lumbricales 3 and 4** occurs so that **flexion of the ring and little fingers at the metacarpophalangeal joint** is lost. However, the extensor muscles are intact so that there is hyperextension at these joints.

4. Paralysis of **palmar and dorsal interosseous muscles** occurs so that **abduction and adduction of the fingers** are lost.

5. Paralysis of the **adductor pollicis muscle** occurs so that **adduction of the thumb** is lost.

6. Paralysis of the **abductor digiti minimi, flexor digiti minimi,** and **opponens digiti minimi muscles** occurs so that **abduction, flexion, and opposition of the little finger** are lost.

7. Sensory loss occurs on the **palmar and dorsal aspects of the little finger and half of the ring finger.**

8. Clinically, **clawhand** is observed, but it is not as severe as with a lesion at the elbow or axilla.

IV. SHOULDER REGION

A. Glenohumeral joint

1. The **rotator cuff** contributes to the **stability of the glenohumeral joint** (along with the **tendon of the long head of the biceps brachii muscle**) by holding the head of the humerus against the glenoid surface of the scapula. The rotator cuff is formed by the tendons of the following muscles:

a. **Subscapularis muscle,** innervated by the subscapularis nerve

b. **Supraspinatus muscle,** innervated by the suprascapular nerve

   c. **Infraspinatus muscle,** innervated by the suprascapular nerve

   d. **Teres minor muscle,** innervated by the axillary nerve

   2. **Dislocation of the humerus (shoulder dislocation)** frequently occurs because of the shallowness of the glenoid fossa.

   a. **Anterior dislocation.** The head of the humerus lies inferior to the **coracoid process** of the scapula and may damage the **axillary nerve.**

   b. **Posterior dislocation.** The head of the humerus is displaced posteriorly.

B. **Acromioclavicular subluxation (shoulder separation).** The acromioclavicular joint is reinforced by the **acromioclavicular** and **coracoclavicular ligaments.** Despite these ligamentous reinforcements, shoulder separation is common and results in a downward displacement of the clavicle.

C. **Fracture of the clavicle.** This fracture most commonly occurs at the middle one-third of the clavicle and results in the upward displacement of the proximal fragment (due to the pull of the sternocleidomastoid muscle) and downward displacement of the distal fragment (due to the pull of the deltoid muscle and gravity).

D. **Subacromial bursitis** and **supraspinatus tendinitis** involves inflammation of the subacromial bursa, which separates the acromion from the supraspinatus muscle.

V. CUBITAL FOSSA CONTENTS (Figure 6-4)

A. Median nerve

B. Brachial artery

C. Biceps tendon

D. Median cubital vein (superficial to the bicipital aponeurosis)

E. Radial nerve (lying deep to the brachioradial muscle)

VI. ELBOW JOINT

A. This joint consists of three articulations among the humerus, ulna, and radius.

   1. The **humeroulnar joint** is reinforced by the **ulnar (medial) collateral ligament.** A sprain of this ligament permits abnormal **abduction of the forearm.**

   2. The **humeroradial joint** is reinforced by the **radial (lateral) collateral ligament.** A sprain of this ligament permits abnormal **adduction of the forearm. Tennis elbow** involves inflammation of this ligament.

   3. The **radioulnar joint** is reinforced by the **annular ligament. Pronation** and **supination** occur at this joint.

B. Pulled elbow

   1. Severe distal traction of the radius (e.g., a parent yanking the arm of a child) can **subluxate the head of the radius** from its articulations with the humerus and ulna and **tear the annular ligament.**

   2. **Pronation** and **supination** will be restricted much more than flexion and extension.

   3. **Reduction of a pulled elbow** involves applying direct pressure posteriorly on the head of the radius, while simultaneously supinating and extending the forearm. This manipulation effectively "screws" the head of the radius into the annular ligament.

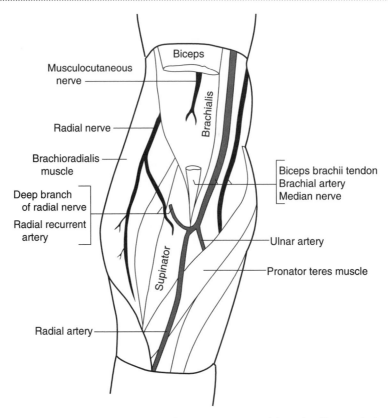

Figure 6-4.   Dissection of the deep structures in the anterior aspect of the right elbow with the cubital fossa open widely. Adapted with permission from Moore, KL: *Clinically Oriented Anatomy*, 3rd edition, Baltimore, Williams & Wilkins, 1992.

VII. WRIST

    A. The flexor retinaculum is attached to the palmar surface of the carpal bones and forms the **carpal tunnel** through which the following structures pass:

        1. Flexor digitorum superficialis tendons

        2. Flexor digitorum profundus tendons

        3. Flexor pollicis longus tendon

        4. Median nerve

    B. **Compression of the median nerve** within the carpal tunnel results in **carpal tunnel syndrome** with the following symptoms:

        1. Flexion and abduction of the thumb are weakened.

        2. Opposition of the thumb is lost.

        3. Extension of the index and middle fingers is lost.

        4. Sensory loss palmar and dorsal aspects of the index, middle, and half of the ring fingers and the palmar aspect of the thumb.

        5. Sensation on the radial side of the palm is preserved because the superficial branch of the median is not involved.

C. A **deep laceration on the radial side** of the wrist, as in a suicide attempt, may cut the following structures:

    1. Radial artery

    2. Median nerve

    3. Flexor carpi radialis tendon

    4. Palmaris longus tendon

D. A **deep laceration on the ulnar side** of the wrist (as in a suicide attempt) may cut the following structures:

    1. Ulnar artery

    2. Ulnar nerve

    3. Flexor carpi ulnaris tendon

VIII. FRACTURES (Figure 6-5)

A. Greater tuberosity

    1. This fracture may be associated with a **shoulder separation.**

    2. The **supraspinatus, infraspinatus,** and **teres minor muscles** (muscles of the rotator cuff) draw the greater tuberosity in a superior–posterior direction.

B. Lesser tuberosity. The **subscapularis muscle** draws the lesser tuberosity medially.

C. Surgical neck of the humerus

    1. The **pectoralis major, latissimus dorsi,** and **teres major muscles** draw the distal fragment of the humerus medially.

    2. This fracture may injure the **axillary nerve.**

D. Humerus proximal to the insertion of the deltoid muscle

    1. The **pectoralis major, latissimus dorsi,** and **teres major muscles** draw the proximal fragment medially.

    2. The **deltoid, biceps brachii,** and **triceps muscles** pull the distal fragment proximally.

E. Humerus distal to the insertion of the deltoid muscle

    1. The **deltoid muscle** draws the proximal fragment laterally.

    2. The **biceps brachii** and **triceps muscles** pull the distal fragment proximally.

    3. This fracture may injure the **radial nerve.**

F. Supracondylar fracture of the humerus

    1. The **triceps brachii muscle** draws the ulna posteriorly.

    2. The **common flexor tendon** of the forearm draws the distal fragment of the humerus distally.

    3. This fracture may injure the **brachial artery** and **median nerve.**

G. Medial epicondyle

    1. If the forearm is forcibly abducted, the **medial collateral ligament** of the **elbow joint** can avulse the medial epicondyle.

    2. The **common flexor tendon** of the forearm draws the medial epicondyle distally.

    3. This fracture may injure the **ulnar nerve.**

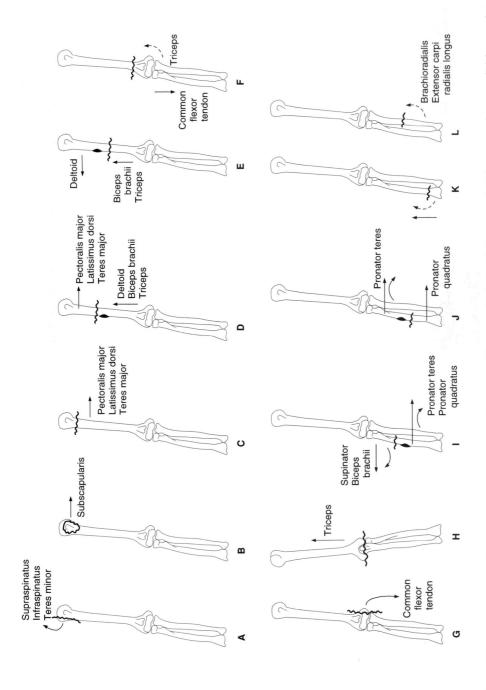

Figure 6-5. A schematic diagram depicting various bony fractures of the upper limb (A–L). Fracture of: (A) greater tuberosity; (B) lesser tuberosity; (C) surgical neck of humerus; (D) humerus proximal to insertion of deltoid; (E) humerus distal to insertion of deltoid; (F) supracondylar fracture of humerus; (G) medial epicondyle of humerus; (H) olecranon process; (I) radius proximal to insertion of pronator teres; (J) radius distal to insertion of pronator teres; (K) distal radius (Colles'); (L) ulna. Arrows indicate the direction of displacement of the bony fragments. The muscles responsible for the displacements are indicated. Dotted arrows indicate a posterior displacement. ⌣ supination. ⌢ pronation.

H. Olecranon process. The **triceps brachii muscle** draws the proximal fragment proximally.

I. Radius proximal to the insertion of pronator teres muscle

    1. The **supinator** and **biceps brachii muscles** supinate the proximal fragment and draw it laterally.

    2. The **pronator teres** and **pronator quadratus muscles** pronate the distal fragment and draw it medially.

J. Radius distal to the insertion of pronator teres muscle

    1. The **pronator teres muscle** pronates the proximal fragment and draws it medially.

    2. The **pronator quadratus muscle** draws the distal fragment medially.

K. Distal radius (Colles' fracture)

    1. The distal fragment is displaced in a posterior and superior direction, thus shortening the radius.

    2. This fracture is often accompanied by a fracture of the styloid process of the ulna.

L. Shaft of the ulna. The **brachioradialis** and **extensor carpi radialis longus muscles** draw the distal fragment posteriorly.

M. Scaphoid. Usually there is very little displacement. This fracture is **easily missed** on radiographs and can be misdiagnosed as a sprain. If a radiograph is repeated 10 days after the injury when the bone resorption phase of healing has occurred, the diagnosis will be apparent. Fracture of the scaphoid may interrupt the blood supply to a portion of the bone and cause avascular necrosis.

IX. CROSS-SECTIONAL ANATOMY

A. Cross section of the right arm (Figure 6-6)

B. Cross section of the right forearm (Figure 6-7)

X. RADIOLOGY

A. Anteroposterior (AP) radiograph of the shoulder (Figure 6-8)

B. AP radiograph of the elbow (Figure 6-9)

C. AP radiograph of the hand (Figure 6-10)

D. T1 magnetic resonance image of the shoulder (Figure 6-11)

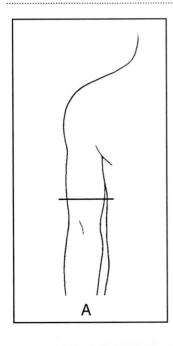

KEY
ABL abductor pollicis longus
AIA anterior interosseous
  artery
BB biceps brachii
BD brachioradialis
BR brachialis
BV basilic vein
CV cephalic vein
ECB extensor carpi radialis brevis
ECL extensor carpi radialis longus
ECU extensor carpi ulnaris
ED extensor digitorum
EDM extensor digiti minimi
EPB extensor pollicis brevis
EPL extensor pollicis longus
FCR flexor carpi radialis
FCU flexor carpi ulnaris
FDP flexor digitorum profundus
FDS flexor digitorum superficialis
FPL flexor pollicis longus

FT subcutaneous fat
H shaft of humerus
IOM interosseous membrane
LOH long head triceps
LS lateral intermuscular septum
LTH lateral head triceps
MCN musculocutaneous nerve
MH medial head triceps
MN median nerve
MS medial intermuscular septum
PB profunda brachii artery
PIA posterior interosseous artery
PL palmaris longus
R radius
RA radial artery
RN radial nerve
SRN superficial radial nerve
TCB triceps brachii
U ulna
UA ulnar artery
UN ulnar nerve

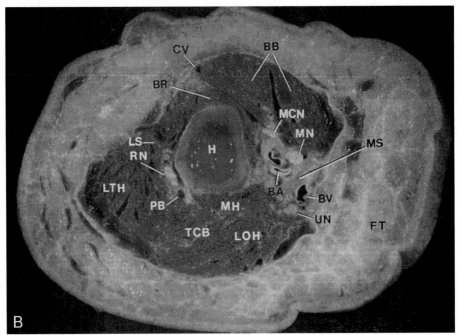

Figure 6-6.   A cadaver cross section through the right arm. Reproduced with permission from Barrett, CP ; Poliakoff, SJ; Holder, LE; et al.: *Primer of Sectional Anatomy with MRI and CT Correlation,* 2nd edition, Baltimore, Williams & Wilkins, 1994.

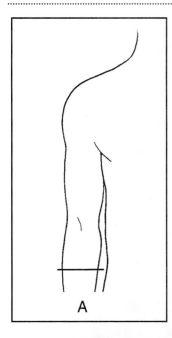

KEY
ABL abductor pollicis longus
AIA anterior interosseous artery
BB biceps brachii
BD brachioradialis
BR brachialis
BV basilic vein
CV cephalic vein
ECB extensor carpi radialis brevis
ECL extensor carpi radialis longus
ECU extensor carpi ulnaris
ED extensor digitorum
EDM extensor digiti minimi
EPB extensor pollicis brevis
EPL extensor pollicis longus
FCR flexor carpi radialis
FCU flexor carpi ulnaris
FDP flexor digitorum profundus
FDS flexor digitorum superficialis
FPL flexor pollicis longus
FT subcutaneous fat

H shart of humerus
IOM interosseous membrane
LOH long head triceps
LS lateral intermuscular septum
LTH lateral head triceps
MCN musculocutaneous nerve
MH medial head triceps
MN median nerve
MS medial intermuscular septum
PB profunda brachii artery
PIA posterior interosseous artery
PL palmaris longus
R radius
RA radial artery
RN radial nerve
SRN superficial radial nerve
TCB triceps brachii
U ulna
UA ulnar artery
UN ulnar nerve

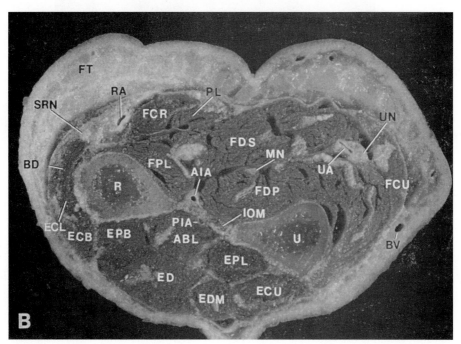

Figure 6-7. A cadaver cross section through the right forearm. Reproduced with permission from Barrett, CP ; Poliakoff, SJ; Holder, LE; et al.: *Primer of Sectional Anatomy with MRI and CT Correlation*, 2nd edition, Baltimore, Williams & Wilkins, 1994.

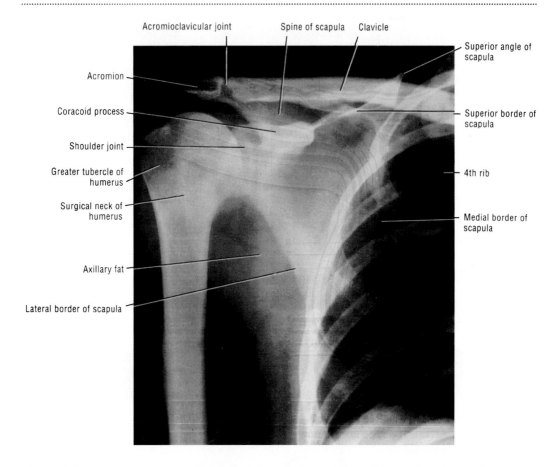

Figure 6-8.   An anteroposterior radiograph of the right shoulder. Reproduced with permission from Moore, KL: *Clinically Oriented Anatomy,* 3rd edition, Baltimore, Williams & Wilkins, 1992.

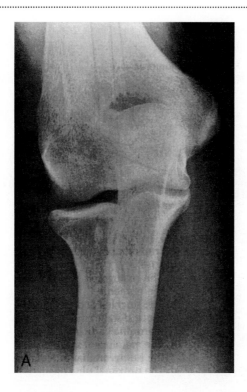

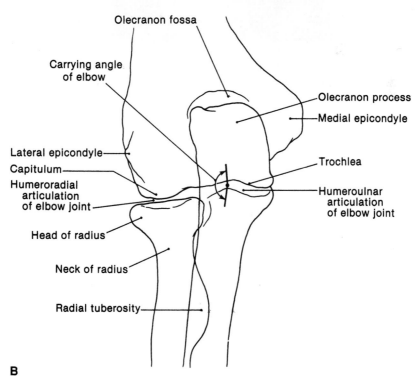

Figure 6-9.   (A) An anteroposterior radiograph of the right elbow and (B) its schematic representation. Reproduced with permission from Slaby, F and Jacobs, ER: *Radiographic Anatomy*, Baltimore, Williams & Wilkins, 1990.

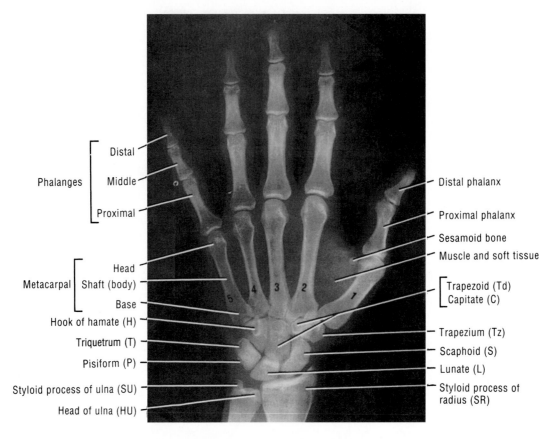

Figure 6-10.   An anteroposterior radiograph of the right hand. Reproduced with permission from Moore, KL: *Clinically Oriented Anatomy*, 3rd edition, Baltimore, Williams & Wilkins, 1992.

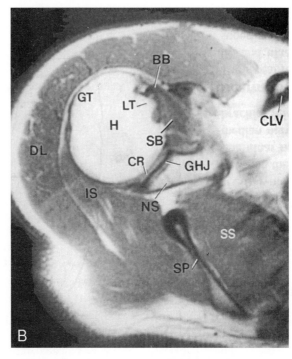

KEY

AP acromion process	**GL** superior glenoid labrum	**PH** epiphyseal line
BB tendon of biceps brachii (long head)	**GP** glenoid process	**SB** subscapularis
CLV clavicle	**GT** greater tubercle of humerus	**SBT** subscapularis tendon
CR hyaline articular cartilage	**H** head of humerus	**SDB** subdeltoid bursa
DL deltoid	**IS** infraspinatus	**SP** body of scapula
GF glenoid fossa	**LT** lesser tubercle of humerus	**SS** supraspinatus
GHJ glenohumeral joint	**N** neck of humerus	**TZ** trapezius
	NS neck of scapula	

Figure 6-11.   A T1 magnetic resonance image of the right shoulder in the transverse plane. Note that three muscles of the rotator cuff are shown (subscapularis, infraspinatus, supraspinatus). The teres minor muscle is not in the plane of section. The subscapularis muscle inserts on the lesser tuberosity of the humerus. All other rotator cuff muscles insert on the greater tuberosity of the humerus. Micrograph reproduced with permission from Barrett, CP ; Poliakoff, SJ; Holder, LE; et al.: *Primer of Sectional Anatomy with MRI and CT Correlation*, 2nd edition, Baltimore, Williams & Wilkins, 1994.

7

Lower Limb

I. ARTERIAL SUPPLY (Figure 7-1)

A. The **obturator artery** is a continuation of the internal iliac artery and passes through the obturator foramen. The obturator artery branches off into the following structures:

1. Muscular branches to the adductor muscles

2. Artery to the head of the femur

B. The **femoral artery** is a continuation of the external iliac artery and enters the thigh posterior to the inguinal ligament and midway between the anterior-superior iliac spine and the symphysis pubis where it can be palpated. The femoral artery branches off into the following arteries:

1. Superficial epigastric artery

2. Superficial circumflex iliac artery

3. External pudendal artery

4. The **profunda femoris artery** branches off into the following arteries:

    a. Four perforating arteries

    b. Medial circumflex artery

    c. Lateral circumflex artery

C. The **popliteal artery** is a continuation of the femoral artery at the adductor hiatus in the adductor magnus muscle and extends through the popliteal fossa where it can be palpated. It branches off into the following arteries:

1. Genicular arteries

2. The **anterior tibial artery** descends on the anterior surface of the interosseous membrane with the deep peroneal nerve and terminates as the **dorsalis pedis artery.** The dorsalis pedis artery lies between the extensor hallucis longus and extensor digitorum longus tendons midway between the medial and lateral malleolus where it can be palpated.

3. The **posterior tibial artery** passes behind the medial malleolus with the tibial nerve where it can be palpated. The posterior tibial artery branches into the following arteries:

    a. Peroneal artery; passes behind the lateral malleolus.

    b. Medial plantar artery

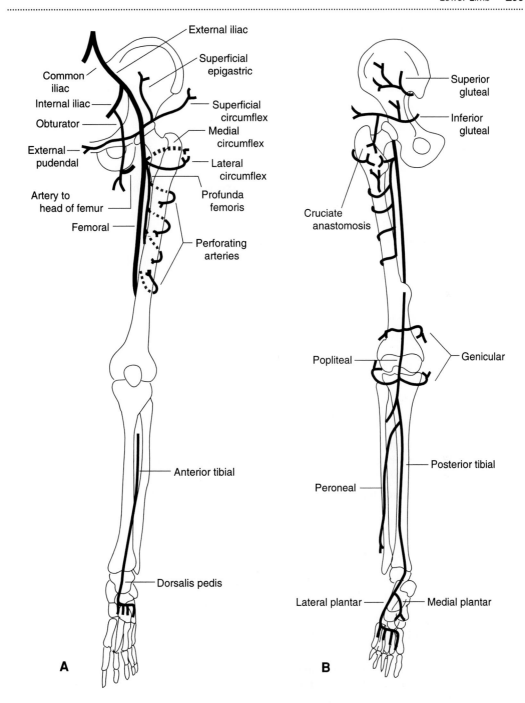

Figure 7-1. A schematic diagram of the arterial supply of the lower limb. (A) Anterior view; (B) posterior view.

     c. Lateral plantar artery; forms the **plantar arch,** which connects to the dorsalis pedis artery.

D. Collateral circulation

    1. **Around the hip joint (cruciate anastomosis)** involves the following arteries:

        a. **Inferior gluteal artery** (a branch of the internal iliac artery)

        b. **Medial femoral circumflex artery**

        c. **Lateral femoral circumflex artery**

        d. **First perforating branch of profundus femoris artery**

    2. **Around the head of the femur (trochanteric anastomosis)** involves the following arteries:

        a. **Superior gluteal artery**

        b. **Inferior gluteal artery**

        c. **Medial femoral circumflex artery,** which is a **major supply** of blood to the head and neck of the femur

        d. **Lateral femoral circumflex artery**

        e. **Artery to the head of the femur**

            (1) Although not considered a part of the trochanteric anastomosis, this artery is of considerable importance in **children** because it supplies the head of the femur **proximal** to the epiphyseal growth plate.

            (2) After the epiphyseal growth plate closes in the **adult,** this artery plays an insignificant role in supplying blood to the head of the femur.

E. Placement of ligatures

    1. In **emergency situations,** the femoral artery can be ligated anywhere along its course in the anterior compartment of the thigh without risking total loss of blood supply to the lower limb distal to the ligature site.

    2. However, **sudden occlusion** of the femoral artery by ligature or embolism is usually followed by **gangrene.** In general, collateral circulation in the lower limb is not as robust as in the upper limb.

F. **Atherosclerotic occlusive disease of the leg** causes muscular pain or fatigue that occurs with exercise (e.g., walking) but subsides with rest. This is known as **intermittent claudication.** The occlusion may occur in the following arteries:

    1. The **external iliac artery** diminishes blood supply to almost all of the muscles of the lower limb and causes pain in the gluteal region, thigh, and leg.

    2. The **femoral artery proximal to the profunda femoris artery** diminishes blood supply to the thigh and leg muscles and causes pain in the thigh and leg.

    3. The **femoral artery distal to the profunda femoris artery** diminishes blood supply to the leg muscles and causes pain in the leg.

    4. The **popliteal artery** diminishes blood supply to the leg muscles and causes pain in the leg.

    5. The **anterior** or **posterior tibial artery** does **not** diminish blood supply to the leg muscle and, therefore, no intermittent claudication is apparent.

II. **LUMBOSACRAL PLEXUS.** The subdivisions of the lumbosacral plexus include:

   A. Rami are the **L1–L4** and **S1–S4 ventral primary rami** of spinal nerves.

   B. **Divisions (anterior and posterior)** are formed by rami dividing into anterior and posterior divisions.

   C. **Branches.** The six major terminal branches are:

     1. Superior gluteal nerve

     2. Inferior gluteal nerve

     3. Obturator nerve

     4. Femoral nerve

     5. **Tibial nerve.** The tibial nerve and common peroneal nerve comprise the **sciatic nerve.**

     6. **Common peroneal nerve** divides into the:

       a. Superficial peroneal nerve

       b. Deep peroneal nerve

III. **NERVE LESIONS**

   A. **Superior gluteal nerve injury** may be caused during surgery or as a result of poliomyelitis.

     1. Paralysis of **gluteus medius** and **gluteus minimus muscles** occurs so that the **ability to pull the pelvis down** and **abduction of the thigh** is lost.

     2. Clinically, this condition is called **gluteus medius limp** or **waddling gait.** The patient will demonstrate a positive **Trendelenburg's sign,** which is tested as follows:

       a. The patient stands with his back to the examiner and alternatively raises each foot off the ground. If the superior gluteal nerve on the **left** side is injured, then when the patient raises his right foot off the ground, the **right** pelvis will fall downward. Note that it is the side **contralateral** to the nerve injury that is affected.

       b. A Trendelenburg sign can also be observed in a patient with a **hip dislocation** or **fracture of the neck of the femur.**

   B. **Inferior gluteal nerve injury** may be caused during surgery.

     1. Paralysis of the **gluteus maximus muscle** occurs so that the ability to **rise from a seated position, climb stairs,** or **jump** is lost.

     2. Clinically, the patient will be able to walk. However, the patient will **lean the body trunk backward at heel strike** to compensate for the loss of gluteus maximus function.

   C. **Obturator nerve injury** is rare.

     1. Paralysis of a portion of the **adductor magnus, adductor longus,** and **adductor brevis muscles** occur so that **adduction of the thigh** is lost.

     2. Sensory loss occurs on the **medial aspect of the thigh.**

   D. **Femoral nerve injury** may be caused by trauma of the femoral triangle.

     1. Paralysis of the **iliacus** and **sartorius muscles** occurs so that **flexion of the thigh** is weakened.

     2. Paralysis of the **quadriceps femoris muscles** occurs so the **extension of the leg** is lost.

     3. Sensory loss occurs on the **anterior aspect of the thigh** and **medial aspect of the leg and foot.**

E. Tibial nerve injury (at the popliteal fossa) may be caused by trauma of the popliteal fossa.

   1. Paralysis of the **gastrocnemius, soleus,** and **plantaris muscles** occurs so that **plantar flexion of the foot** is lost.

   2. Paralysis of the **flexor digitorum longus** and **flexor hallucis longus muscles** occurs so that **flexion of the toes** is lost.

   3. Paralysis of **tibialis posterior muscle** occurs so that **inversion of the foot** is weakened.

   4. Sensory loss occurs on the **sole of the foot.**

   5. Clinically, the patient will present with **calcaneovalgocavus** in which the opposing muscles will cause **dorsiflexion and eversion of the foot.**

F. Common peroneal nerve injury may be caused by a blow to the lateral aspect of the leg or fracture of the neck of the fibula. This is a common type of injury.

   1. Paralysis of **peroneus longus** and **peroneus brevis muscles** occurs so that **eversion of the foot** is lost.

   2. Paralysis of **tibialis anterior muscle** occurs so that **dorsiflexion of the foot** is lost.

   3. Paralysis of **extensor digitorum longus** and **extensor hallucis longus** occurs so that **extension of the toes** is lost.

   4. Sensory loss occurs on the **anterolateral aspect of the leg** and **dorsum of the foot.**

   5. Clinically, the patient will present with **equinovarus** in which the opposing muscles will cause **plantar flexion (footdrop)** and **inversion of the foot.** The patient has a high stepping gait in which the foot is raised higher than normal so that the toes do not hit the ground. In addition, the foot is brought down suddenly, producing a "clopping" sound.

IV. GLUTEAL REGION

A. The **piriformis muscle** is the **landmark** of the gluteal region. The superior gluteal vessels and nerve emerge superior to the piriformis muscle, whereas the inferior gluteal vessels and nerve emerge inferior to it.

B. Gluteal intramuscular injections can be made safely in the **superolateral** portion of the buttock.

C. Hip joint

   1. Ligamentous support

      a. Iliofemoral ligament (Y ligament of Bigelow)

      b. Pubofemoral ligament

      c. Ischiofemoral ligament

      d. Ligamentum teres plays only a minor role in stability but does carry the **artery to the head of the femur.**

   2. Posterior dislocation of the hip joint

      a. Because the hip joint is very stable, a dislocation usually is caused by a severe trauma (e.g., car accident) that fractures the acetabulum.

      b. The head of the femur comes to lie posterior to the iliofemoral ligament.

      c. If the articular capsule is damaged, blood supply to the head of the femur is in jeopardy.

d. The sciatic nerve may be damaged.

e. Clinically, the patient will present with a lower limb that is **flexed at the hip joint, adducted, medially rotated,** and **shorter** than the opposite limb. Note the difference in how the lower limb presents in a posterior dislocation versus a **fracture of the neck of the femur** (see X A).

V. FEMORAL TRIANGLE (Figure 7-2) contains the following structures:

A. Femoral canal, containing lymphatics and lymph nodes

B. Femoral vein. The **great saphenous vein** joins the femoral vein within the femoral triangle just below and lateral to the pubic tubercle. This is an important site where a great saphenous vein cutdown can be performed.

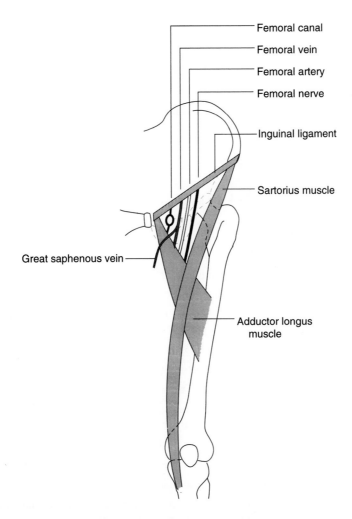

Figure 7-2. A schematic diagram of the femoral triangle. Note the great saphenous vein joins the femoral vein at the femoral triangle. The boundaries of the femoral triangle are the inguinal ligament, sartorius muscle, and adductor longus muscle. The most lateral structure in the femoral triangle is the femoral nerve.

C. Femoral artery

D. Femoral nerve

VI. POPLITEAL FOSSA contains the following structures:

A. Tibial nerve

B. Common peroneal nerve

C. Popliteal artery

D. Popliteal vein

E. Small saphenous vein

VII. KNEE JOINT

A. Menisci

1. Medial meniscus is a C-shaped fibrocartilage that is attached to the tibial collateral ligament.

2. Lateral meniscus is an O-shaped fibrocartilage.

B. Ligaments

1. Patellar ligament is struck to elicit the knee-jerk reflex. The reflex is blocked by damage to the femoral nerve, which supplies the quadriceps muscle, or damage to spinal cord segments L2, 3, 4.

2. Tibial (medial) collateral ligament

a. Extends from the medial epicondyle of the femur to the shaft of the tibia

b. Prevents **abduction** of the knee joint

c. A torn tibial collateral ligament can be recognized by abnormal passive abduction of the extended leg.

3. Fibular (lateral) collateral ligament

a. Extends from the lateral epicondyle of the femur to the head of the fibula

b. Prevents **adduction** of the knee joint

c. A torn fibular collateral ligament can be recognized by the abnormal passive adduction of the extended leg.

4. Anterior cruciate ligament (Figure 7-3)

a. Extends from the **anterior** aspect of the tibia to the lateral condyle of the femur

b. Prevents **anterior** movement of the tibia in reference to the femur

c. A torn anterior cruciate ligament can be recognized by abnormal passive **anterior** displacement of the tibia called an **anterior drawer sign.**

5. Posterior cruciate ligament (see Figure 7-3)

a. Extends from the **posterior** aspect of the tibia to the medial condyle of the femur

b. Prevents **posterior** movement of the tibia in reference to the femur

c. A torn posterior cruciate ligament can be recognized by abnormal passive **posterior** displacement of the tibia called a **posterior drawer sign.**

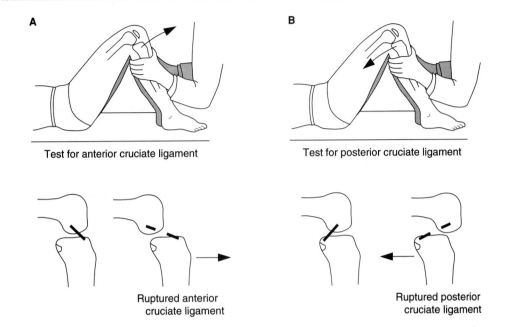

Figure 7-3. A diagram depicting the clinical test for (A) a ruptured anterior cruciate ligament (anterior drawer sign) and (B) a ruptured posterior cruciate ligament (posterior drawer sign). Adapted with permission from Snell, RS: *Clinical Anatomy for Medical Students,* 5th edition, Boston, Little, Brown and Company (Inc.), 1992.

C. Combined knee injury ("terrible triad") involves the following structures:

1. **Tibial collateral ligament,** which is torn due to excessive abduction of the knee joint

2. **Medial meniscus,** which is torn as a result of its attachment to the tibial collateral ligament

3. **Anterior cruciate ligament** is torn

VIII. ANKLE (TALOCRURAL) JOINT. This joint involves **dorsiflexion** and **plantar flexion of the foot.**

A. Ligaments

1. **Medial (deltoid) ligament**

a. Extends from the medial malleolus to the tarsal bones

b. Prevents **abduction** of the ankle joint and limits dorsiflexion and plantar flexion

2. **Lateral ligaments**

a. Extend from the lateral malleolus to the talus and calcaneus

b. Prevent **adduction** of the ankle joint and limit dorsiflexion and plantar flexion

B. Ankle injuries

1. **Pott's fracture**

a. This fracture occurs when the foot is forcibly **everted,** which pulls the **medial ligament.**

    b. The medial ligament is so strong that, instead of tearing, it causes an **avulsion of the medial malleolus.**

    c. Subsequently, the talus moves laterally and causes a **fracture of the fibula.**

  2. Inversion sprain

    a. This sprain is the **most common ankle injury** and occurs when the foot is forcibly **inverted,** which **stretches or tears the lateral ligaments.**

    b. Avulsion of the lateral malleolus or **avulsion of the tuberosity of the fifth metatarsal** (where the **peroneus brevis muscle** attaches) may occur, depending on the severity of the injury.

C. **Medial malleolus** has several important related structures.

  1. Anterior relationships

    a. Saphenous nerve

    b. Great saphenous vein. This is an excellent location for a great saphenous vein cutdown.

  2. Posterior relationships

    a. Flexor hallucis longus tendon

    b. Flexor digitorum longus tendon

    c. Tibial posterior tendon

    d. Posterior tibial artery

    e. Tibial nerve

IX. SUBTALAR (TALOCALCANEAN) AND TRANSVERSE TARSAL JOINTS. These joints involve **inversion** and **eversion** of the foot.

X. FRACTURES (Figure 7-4)

A. Neck of the femur

  1. The **gluteus maximus, piriformis, obturator internus, gemelli,** and **quadratus femoris muscles** rotate the distal fragment laterally.

  2. The **rectus femoris, adductor,** and **hamstring muscles** draw the distal fragment proximally.

  3. Thus, the lower limb presents as **laterally rotated** and **shortened.**

B. Upper one-third of the femur shaft

  1. The **iliopsoas muscle** flexes the proximal fragment.

  2. The **gluteus medius** and **minimus muscles** draw the proximal fragment laterally.

  3. The **gluteus maximus, piriformis, obturator internus, gemelli,** and **quadratus femoris muscles** rotate the proximal fragment laterally.

  4. The **adductor muscles** draw the distal fragment medially and rotate it laterally.

  5. The **quadriceps femoris** and **hamstring muscles** draw the distal fragment proximally.

C. Middle one-third of the femur shaft

  1. The **quadriceps femoris** and **hamstring muscles** draw the distal fragment proximally.

  2. The **gastrocnemius muscle** draws the distal fragment posteriorly.

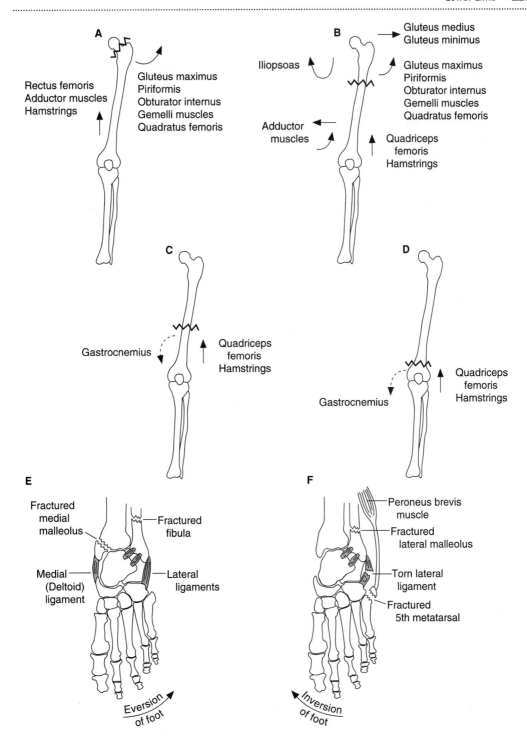

Figure 7-4. A schematic diagram depicting various bony fractures of the lower limb (A–F). Fracture of the (A) neck of the femur; (B) upper one-third of the femur; (C) middle one-third of the femur; (D) distal one-third of the femur; (E) Pott's fracture; (F) inversion sprains. Arrows indicate the direction of displacement of the bony fragments. The muscles responsible for the displacement are indicated. (⤴) Lateral rotation; (↳) Flexion; (⤏) Posterior movement.

   D. Distal one-third of the femur shaft

       1. The **quadriceps femoris** and **hamstring muscles** draw the distal fragment proximally.

       2. The **gastrocnemius muscle** draws the distal fragment posteriorly to a great degree and may interfere with blood flow through the popliteal artery.

   E. **Pott's fracture** (see VIII B 1)

   F. **Inversion sprains** (see VIII B 2)

   G. **Tibia and fibula.** If only one bone is broken, the other bone acts as a splint so that there is little displacement.

XI. CROSS-SECTIONAL ANATOMY

   A. **Cross section of the left thigh** (Figure 7-5)

   B. **Cross section of the left leg** (Figure 7-6)

XII. RADIOLOGY

   A. **Anteroposterior (AP) radiograph of the knee** (Figure 7-7)

   B. **Coronal magnetic resonance image of the knee** (Figure 7-8)

   C. **AP radiograph of the ankle** (Figure 7-9)

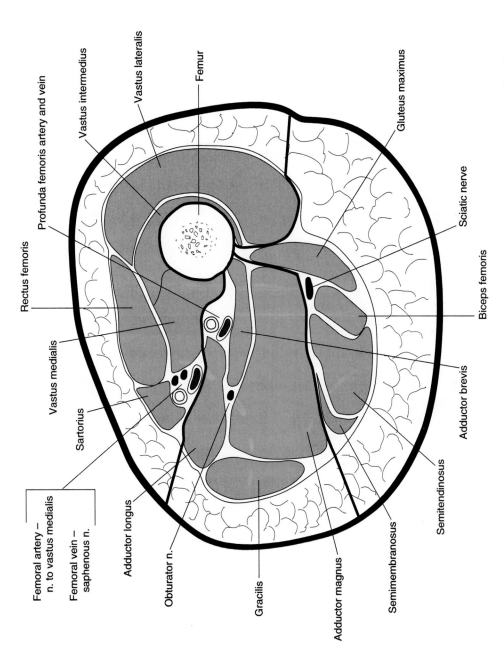

Figure 7-5. A schematic diagram of a cross section through the left thigh. Adapted with permission from Moore, KL: *Clinically Oriented Anatomy*, 3rd edition, Baltimore, Williams & Wilkins, 1992.

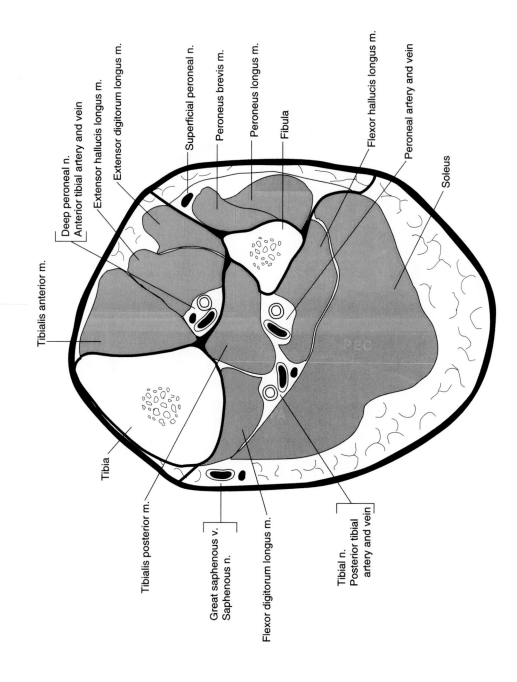

Figure 7-6. A schematic diagram of a cross section through the left leg. Adapted with permission from Moore, KL: *Clinically Oriented Anatomy*, 3rd edition, Baltimore, Williams & Wilkins, 1992.

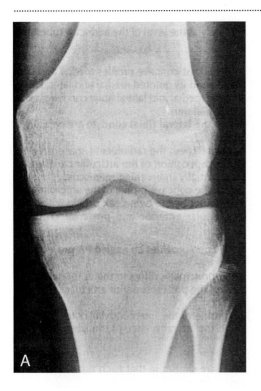

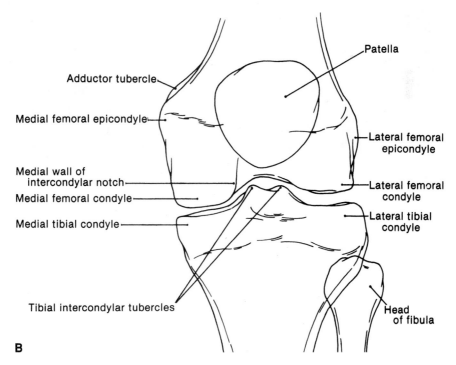

Figure 7-7. (A) An anteroposterior (AP) radiograph of the left knee. (B) A schematic representation. Reproduced with permission from Slaby, F and Jacobs, ER: *Radiographic Anatomy,* Baltimore, Williams & Wilkins, 1990.

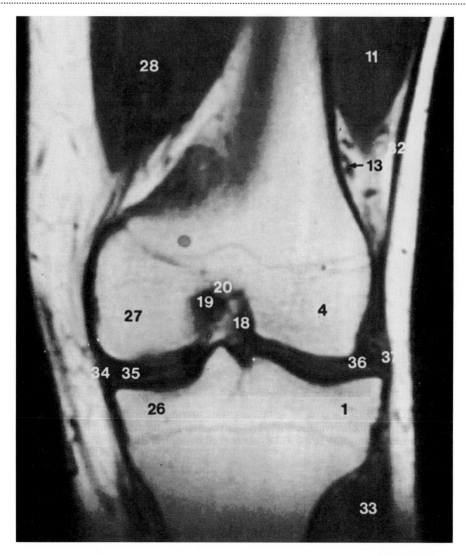

Figure 7-8. A coronal T1 magnetic resonance image of the left knee. (*1*) Lateral condyle of tibia; (*4*) lateral condyle of femur; (*11*) vastus lateralis muscle; (*13*) superior lateral genicular artery; (*18*) anterior cruciate ligament; (*19*) posterior cruciate ligament; (*20*) intercondylar notch; (*26*) medial condyle of the tibia; (*27*) medial condyle of femur; (*28*) vastus medialis muscle; (*32*) iliotibial tract; (*33*) tibialis anterior muscle; (*34*) tibial (medial) collateral ligament; (*35*) medial meniscus; (*36*) lateral meniscus; (*37*) fibular (lateral) collateral ligament. Reproduced with permission from Weir and Abrahams: *Imaging Atlas of Human Anatomy*, London, UK, Mosby International, 1992.

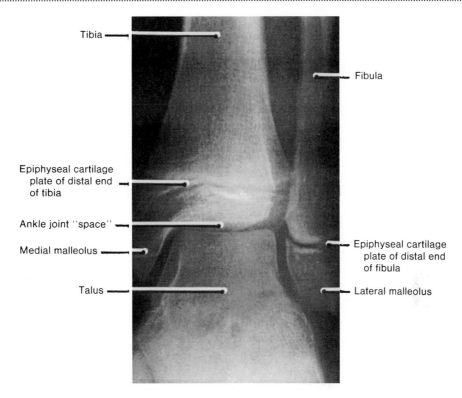

Tibia

Fibula

Epiphyseal cartilage plate of distal end of tibia

Ankle joint ''space''

Medial malleolus

Epiphyseal cartilage plate of distal end of fibula

Talus

Lateral malleolus

Figure 7-9.   An anteroposterior (AP) radiograph of the left ankle joint of a 14-year-old boy. Note how the body of the talus fits into the mortise (slot) formed by the medial and lateral malleoli. Because this is a radiograph of a young boy, the epiphyseal growth plates are still apparent on the radiograph. Reproduced with permission from Weir and Abrahams: *Imaging Atlas of Human Anatomy*, London, UK, Mosby International, 1992.

8

Head and Neck

I. ARTERIAL SUPPLY (Figure 8-1)

A. Branches of the arch of the aorta (see Chapter 2, Figure 2-5)

1. Brachiocephalic artery

a. Right subclavian artery

b. Right common carotid artery

2. Left common carotid artery

3. Left subclavian artery

B. Common carotid artery

1. The **internal carotid artery** has no branches in the neck, but has **three branches** within the cranium.

a. Ophthalmic artery

b. Anterior cerebral artery

c. Middle cerebral artery

2. The **external carotid artery** has eight branches in the neck.

a. Superior thyroid artery

b. Lingual artery

c. Facial artery

d. Ascending pharyngeal artery

e. Occipital artery

f. Posterior auricular artery

g. Superficial temporal artery

h. The **maxillary artery** enters the infratemporal fossa by passing posterior to the neck of the mandible and branches into the following arteries:

(1) The **middle meningeal artery** supplies the periosteal dura in the cranium. Skull fractures in the area of the **pterion** (junction of the parietal, frontal, temporal, and sphenoid bones) may sever the middle meningeal artery, resulting in an **epidural hemorrhage** (see VIII A).

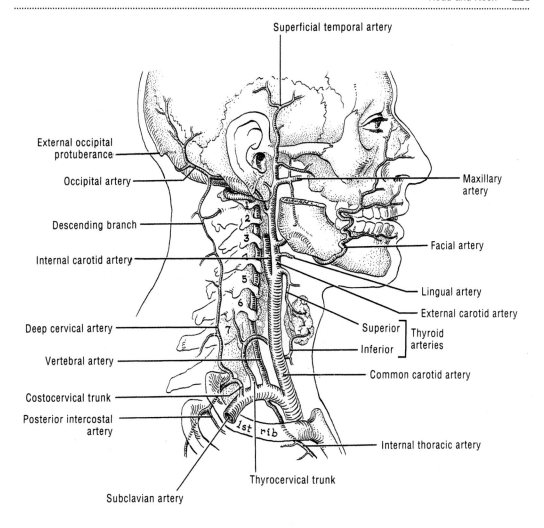

Superficial temporal artery

External occipital protuberance

Occipital artery

Descending branch

Internal carotid artery

Deep cervical artery

Vertebral artery

Costocervical trunk

Posterior intercostal artery

Subclavian artery

Thyrocervical trunk

Maxillary artery

Facial artery

Lingual artery

External carotid artery

Superior | Thyroid
Inferior | arteries

Common carotid artery

Internal thoracic artery

1st rib

Figure 8-1. A diagram of the arterial supply of the head and neck region. Reproduced with permission from Moore, KL: *Clinically Oriented Anatomy*, 3rd edition, Baltimore, Williams & Wilkins, 1992.

> **(2)** Inferior alveolar artery
>
> **(3)** Posterior superior alveolar artery
>
> **(4)** Descending palatine artery
>
> **(5)** Sphenopalatine artery

II. VENOUS DRAINAGE

A. The **facial vein** (no valves) drains into the **internal jugular vein** and provides the major venous drainage of the face.

B. The facial vein makes clinically important connections with the **cavernous sinus** via the **inferior ophthalmic vein, superior ophthalmic vein,** and **pterygoid plexus of veins.** This connection with the cavernous sinus provides a potential route of infection from the **superficial face (danger zone** of the face) to the **dural sinuses** within the cranium.

III.   CERVICAL PLEXUS. The cervical plexus is formed by **C1–C4 ventral primary rami** of spinal nerves.

   A. Motor branches

   1. The **superior ramus of the ansa cervicalis (C1)** innervates the **geniohyoid** and **thyrohyoid muscles.**

   2. The **inferior ramus of the ansa cervicalis (C2, C3)** innervates the **sternohyoid, omohyoid,** and **sternothyroid muscles.**

   3. The **phrenic nerve (C3, C4, C5)** innervates the **diaphragm.**

   B. Sensory branches

   1. Lesser occipital nerve

   2. Greater auricular nerve

   3. Transverse cervical nerve

   4. Supraclavicular nerves

   5. An unnamed sensory branch that passes through the **foramen magnum** to innervate the **meninges.**

IV.   CERVICAL TRIANGLES OF THE NECK (Figure 8-2 and Table 8-1). The **sternocleidomastoid muscle** divides the neck into the **anterior** and **posterior triangles,** both of which are further subdivided. The **carotid** and **occipital triangles** contain important anatomical structures.

   A. Clinical aspects of the anterior triangle

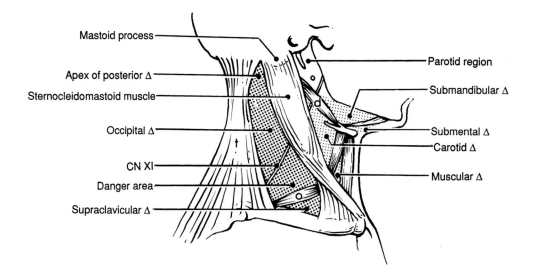

Figure 8-2.   A diagram of the lateral aspect of the neck showing the cervical triangles and their subdivisions. Note CN XI passing through the occipital triangle. Observe that the inferior belly of the omohyoid muscle (O) divides the posterior triangle into the occipital triangle and supraclavicular triangle. The digastric muscle (d) further subdivides the anterior triangle. Reproduced with permission from Moore, KL: *Clinically Oriented Anatomy,* 3rd edition, Baltimore, Williams & Wilkins, 1992.

Table 8-1.

Cervical Triangles of the Neck

Triangle	Boundaries	Contents
Anterior triangle	Sternocleidomastoid muscle Base of mandible Ventral midline	
Digastric	—	—
Submental	—	—
Muscular	—	—
Carotid	—	(Common cartotid artery, internal jugular vein, CN X)*, ansa cervicalis, sympathetic trunk, CN XI, CN XII
Posterior triangle	Sternocleidomastoid muscle Trapezius muscle Clavicle	
Supraclavicular	—	—
Occipital	—	Transverse cervical artery, suprascapular artery, subclavian artery, external jugular vein, brachial plexus, phrenic nerve, lesser occipital nerve, greater auricular nerve, transverse cervical nerve, supraclavicular nerves, CN XI

* All three structures lie within the carotid sheath.

1. The **platysma muscle** lies in the superficial fascia above the anterior triangle and is innervated by the **facial nerve.** Accidental damage during surgery of the facial nerve in this area can result in **distortion of the shape of the mouth.**

2. The **carotid pulse** is easily palpated at the anterior border of the sternocleidomastoid muscle at the level of the superior border of the thyroid cartilage (C5).

3. The **bifurcation of the common carotid artery** into the internal carotid artery and external carotid artery occurs in the anterior triangle of the neck at the level of C4. At the bifurcation, the **carotid body** and **carotid sinus** can be found.

   a. The carotid body is an **oxygen chemoreceptor.** Its sensory information is carried to the central nervous system (CNS) by **CN IX** and **CN X.**

   b. The carotid sinus is a **pressure receptor.** Its sensory information is carried to the CNS by **CN IX** and **CN X.**

4. **Atherosclerosis of the internal carotid artery** within the anterior triangle can result in:

   a. **Visual impairment** in the eye on the **same side** of the occlusion caused by insufficient blood flow to the **retinal artery**

   b. **Motor** and **sensory impairment** of the body on the **opposite side** of the occlusion caused by insufficient blood flow to the **middle cerebral artery**

5. **Internal jugular vein catheterization** can be performed in the anterior triangle by inserting a catheter in an area bounded by the **sternal and clavicular heads of the sternocleidomastoid muscle** and the **medial end of the clavicle.**

B. Clinical aspects of the posterior triangle

    1. Injury to CN XI within the posterior triangle due to surgery or a penetrating wound causes paralysis of the **trapezius muscle** so that **abduction of the arm past the horizontal position** is compromised.

    2. Injuries to the trunks of the brachial plexus that lie in the posterior triangle result in **Erb-Duchenne or Klumpke's syndromes** (see Chapter 6, Table 6-1).

    3. Severe upper limb hemorrhage may be stopped by compressing the subclavian artery against the first rib, applying downward and posterior pressure. The brachial plexus and subclavian artery enter the posterior triangle in an area bounded anteriorly by the **anterior scalene muscle,** posteriorly by the **middle scalene muscle,** and inferiorly by the **first rib.**

V.   LARYNX (Figure 8-3)

    A. Consists of **five major cartilages:**

        1. Cricoid

        2. Thyroid

        3. Epiglottis

        4. Arytenoid, of which there are two

    B. The **ventricle** of the larynx is bounded superiorly by the **vestibular folds (false vocal cords)** and inferiorly by the **vocal folds (true vocal cords).**

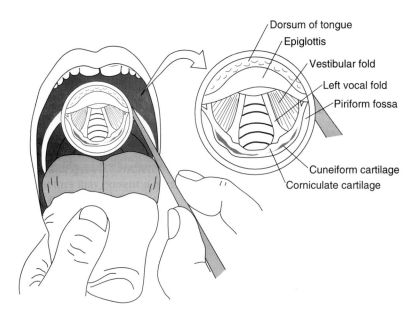

Figure 8-3.   A diagram depicting the anatomical structures observed during inspection of the vocal folds using a laryngeal mirror. Adapted with permission from Pegington, J: *Clinical Anatomy in Action, Volume 2, The Head and Neck,* Edinburgh, UK, Churchill Livingstone, 1986.

C. Intrinsic muscles of the larynx

    1. The **posterior cricoarytenoid muscle** abducts the vocal folds and opens airway during respiration. This is the only muscle that abducts the vocal folds.

    2. The **lateral cricoarytenoid muscle** adducts the vocal folds.

    3. The **arytenoid muscle** adducts the vocal folds.

    4. The **thyroarytenoid muscle** relaxes the vocal folds.

    5. The **vocal muscle** alters vocal folds for speaking and singing.

    6. The **cricothyroid muscle** stretches and tenses vocal folds.

    7. The **transverse** and **oblique arytenoid muscles** close the laryngeal aditus (sphincter function).

D. All **intrinsic muscles** of the larynx are innervated by the **inferior laryngeal nerve of CN X** (a continuation of the **recurrent laryngeal nerve**), except the **cricothyroid muscle,** which is innervated by the **external branch of the superior laryngeal nerve of CN X.**

E. The **recurrent laryngeal nerve** may be damaged during **thyroidectomy surgery.** Damage to the recurrent laryngeal nerve results in much more profound effects on the **abduction** of the vocal folds (i.e., posterior cricoarytenoid muscle) than on adduction of the vocal folds.

    1. Unilateral damage to the recurrent laryngeal nerve results in a hoarse voice, inability to speak for long periods, and movement of the vocal fold on affected side moves toward the midline.

    2. Bilateral damage to the recurrent laryngeal nerve results in acute breathlessness (dyspnea) because both vocal folds move toward the midline and close off the air passage.

VI. CRICOTHYROIDECTOMY (Figure 8-4)

A. This procedure requires that a tube be inserted between the cricoid and thyroid cartilages.

B. The incision made for this procedure passes through the following structures:

    1. Skin

    2. Superficial fascia, avoiding the anterior jugular veins

    3. Investing layer of deep cervical fascia

    4. Pretracheal fascia, avoiding the sternohyoid muscle

    5. Cricothyroid ligament, avoiding the cricothyroid muscle

VII. SKULL AND FORAMINA (Figure 8-5, Table 8-2). Clinical vignette questions concerning fractures of the skull can be handled by knowing the structures transmitted through various foramina in the skull.

VIII. INTRACRANIAL HEMORRHAGE

A. An epidural hemorrhage usually results from a **skull fracture of the temporal or parietal bones** or **near the pterion.**

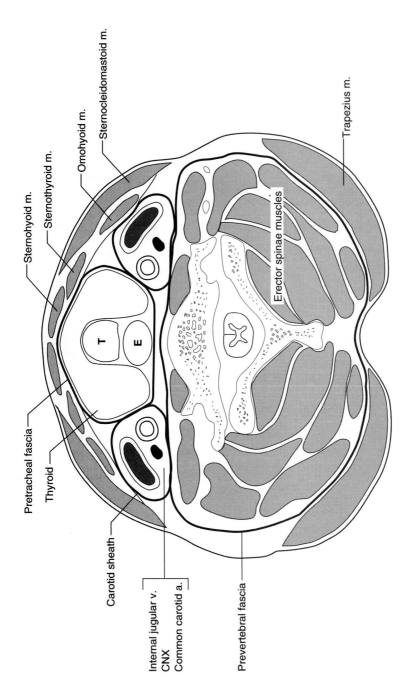

Sternohyoid m.

Sternothyroid m.

Omohyoid m.

Sternocleidomastoid m.

Trapezius m.

Pretracheal fascia

Thyroid

Erector spinae muscles

T

E

Carotid sheath

Internal jugular v.
CNX
Common carotid a.

Prevertebral fascia

Figure 8-4. A schematic diagram of a cross section through the neck. Note the various layers of fascia, the carotid sheath, and the internal jugular vein, CN X, and common carotid artery that lie within the carotid sheath. Study this diagram, paying close attention to the anterior–posterior anatomical relationships. Keep in mind that the USMLE will ask clinical vignette questions concerning in what order the structures will be encountered during, for example, a cricothyroidectomy, other surgical approaches to the neck, or penetrating wounds (e.g., bullet, knife wounds). (*T*) Trachea; (*E*) esophagus.

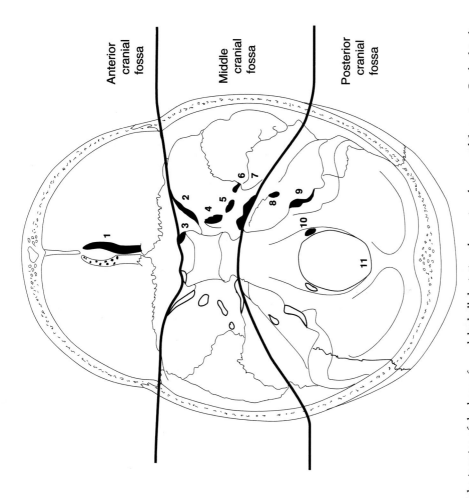

Figure 8-5. A diagram of the interior of the base of an adult skull depicting the various bones and foramina. Study this diagram in conjunction with Table 8-2. Keep in mind that the USMLE will ask clinical vignette questions about fractures to a particular bone or fossa and ask what structures would most likely be affected. 1: Cribriform plate; 2: Superior orbital fissure; 3: Optic canal; 4: Foramen rotundum; 5: Foramen ovale; 6: Foramen spinosum; 7: Foramen lacerum; 8: Internal auditory meatus; 9: Jugular foramen; 10: Hypoglossal canal; 11: Foramen magnum. Adapted with permission from Moore, KL *Clinically Oriented Anatomy*, 3rd edition, Baltimore, Williams & Wilkins, 1992.

Table 8-2.

Skull and Foramina

Opening in Skull	Bone of Skull	Structures Transmitted
Anterior cranial fossa		
Cribiform plate	Ethmoid	CN I
Middle cranial fossa		
Superior orbital fissure	Between lesser and greater wings of sphenoid	CN III, CN IV, CN V_1, CN VI
Optic canal	Lesser wing of sphenoid	CN II
Inferior orbital fissure	Between zygomatic and greater wing of sphenoid	Infraorbital vein, artery, and nerve
Carotid canal	Between petrous part of temporal and greater wing of sphenoid	Internal carotid artery
Foramen rotundurm	Greater wing of sphenoid	CN V_2
Foramen ovale	Greater wing of sphenoid	CN V_3, lesser petrosal nerve
Foramen spinosum	Greater wing of sphenoid	Middle meningeal artery
Foramen lacerum	Between petrous part of temporal and sphenoid	None
Posterior cranial fossa		
Internal acoustic meatus	Petrous part of temporal	CN VII, CN VIII
Jugular foramen	Between petrous part of temporal and occipital	CN IX, CN X, CN XI, sigmoid sinus
Hypoglossal canal	Occipital	CN XII
Foramen magnum	Occipital	Medulla oblongata, CN XI, vertebral arteries

1 Frontal sinus
2 Sagittal suture
3 Crista galli
4 Lambdoid suture
5 Petrous part of temporal bone
6 Internal acoustic meatus
7 Mastoid process
8 Basi-occiput
9 Lateral mass of atlas (first cervical vertebra)
10 Odontoid process (dens) of axis (second cervical vertebra)
11 Floor of maxillary sinus (antrum)
12 Nasal septum
13 Sella turcica
14 Ethmoidal air cells
15 Superior orbital fissure
16 Temporal surface of greater wing
17 Body
18 Lesser wing of sphenoid

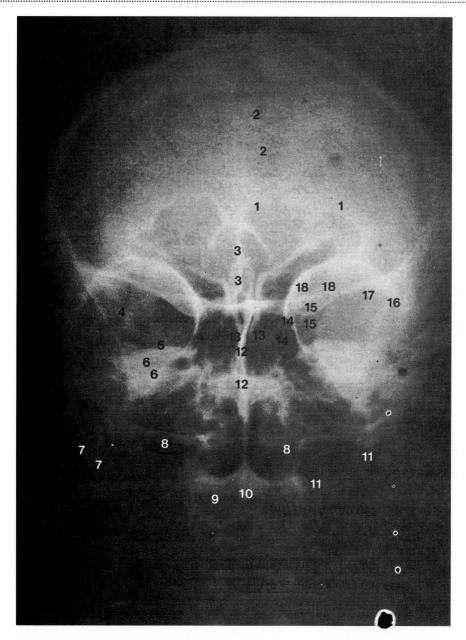

Figure 8-6.   A posteroanterior radiograph of the skull. Note the various numbered structures. Reproduced with permission from Weir and Abrahams: *Imaging Atlas of Human Anatomy*, London, UK, Mosby International, 1992.

1. The fracture tears the **middle meningeal artery,** and blood rapidly accumulates between the internal periosteum of the skull and the outer surface of the dura mater. This situation requires emergency surgery.

2. An epidural hematoma may cause a **transtentorial (uncal) herniation,** which compresses **CN III** and causes a **dilated, fixed pupil.**

3. Blood will **not** appear in the cerebrospinal fluid (CSF).

B. A **subdural hemorrhage** results from either a **blow to the head** or **violent movement** of the brain within the fixed space of the skull.

1. This injury most commonly tears the **superior cerebral veins (bridging veins),** and blood accumulates between the dura and the outer surface of the arachnoid. Note that the blood involved in a subdural hemorrhage is **venous blood.**

2. The bleeding associated with a subdural hemorrhage may be relatively **rapid** (acute) or **gradual** (chronic).

3. Blood will **not** appear in the CSF if the arachnoid is intact.

C. A **subarachnoid hemorrhage** results from either a **contusion/laceration injury** of the brain or an **aneurysm of a cerebral artery** (e.g., Berry aneurysm).

1. In either case, the injury tears a **cerebral artery,** and blood accumulates within the subarachnoid space.

2. Blood will appear in the CSF. Note that the major arteries to the brain travel within the subarachnoid space and, therefore, are surrounded by CSF. A vascular bleed from a major cerebral artery will disseminate within the subarachnoid space, and blood will be found in the CSF if a lumber tap is performed.

3. Blood within the subarachnoid space irritates the meninges and causes a **severe headache, stiff neck,** and **loss of consciousness.**

D. An **intracerebral hemorrhage** results from either **contusion/laceration injury** of the brain or a **penetrating injury** to the brain.

1. The injury tears a **cerebral artery,** and blood accumulates within the brain parenchyma. The arteries most often involved are from the circle of Willis to the basal ganglia and internal capsule and, therefore, cause a **paralytic stroke.**

2. Blood may appear in the CSF.

IX. RADIOLOGY

A. Posteroanterior (PA) radiograph of the skull (Figure 8-6)

B. Lateral radiograph of the skull (Figure 8-7)

C. Coronal magnetic resonance image through the orbit (Figure 8-8)

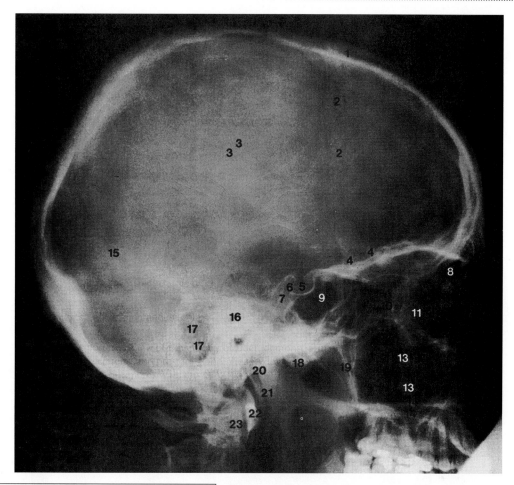

Figure 8-7. A lateral radiograph of the skull. Note the various numbered structures. Reproduced with permission from Weir and Abrahams: *Imaging Atlas of Human Anatomy*, London, UK, Mosby International, 1992.

1 Diploë
2 Coronal suture
3 Grooves for middle meningeal vessels
4 Greater wing of sphenoid
5 Pituitary fossa (sella turcica)
6 Dorsum sellae
7 Clivus
8 Frontal sinus
9 Sphenoidal sinus
10 Ethmoidal air cells
11 Frontal process of zygoma
12 Arch
13 Maxillary process of maxilla
14 Palatine process
15 Lambdoid suture
16 External acoustic meatus
17 Mastoid air cells
18 Articular tubercle for
   temporomandibular joint
19 Coronoid process
20 Condyle of mandible
21 Ramus
22 Anterior arch of atlas (first cervical vertebra)
23 Odontoid process (dens) of axis (second
   cervical vertebra)

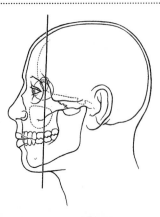

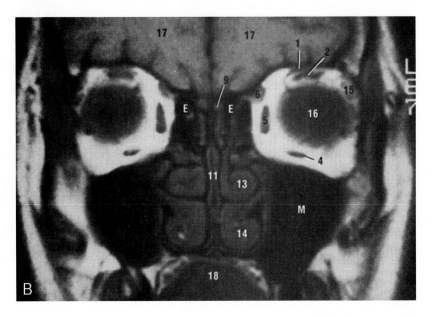

Figure 8-8. A coronal magnetic resonance image through the orbit. (M) Maxillary sinus; (E) ethmoid sinus; (1) levator palpebrae superioris; (2) superior rectus muscle; (4) inferior rectus muscle; (5) medial rectus muscle; (6) superior oblique muscle; (9) olfactory bulb; (11) nasal septum; (13) middle concha; (14) inferior concha; (17) frontal lobe; (18) tongue. Reproduced with permission from Moore, KL: *Clinically Oriented Anatomy*, 3rd edition, Baltimore, Williams & Wilkins, 1992.